EDIBLE MUSHROOMS

EDIBLE MUSHROOMS
Chemical Composition and Nutritional Value

PAVEL KALAČ

Professsor of Agricultural Chemistry,
University of South Bohemia,
České Budějovice, Czech Republic

AMSTERDAM • BOSTON • HEIDELBERG • LONDON
NEW YORK • OXFORD • PARIS • SAN DIEGO
SAN FRANCISCO • SINGAPORE • SYDNEY • TOKYO
Academic Press is an imprint of Elsevier

Academic Press is an imprint of Elsevier
125, London Wall, EC2Y 5AS.
525 B Street, Suite 1800, San Diego, CA 92101-4495, USA
50 Hampshire St., 5th Floor, Cambridge, MA 02139, USA
The Boulevard, Langford Lane, Kidlington, Oxford OX5 1GB, UK

Notices
Knowledge and best practice in this field are constantly changing. As new research
and experience broaden our understanding, changes in research methods, professional
practices, or medical treatment may become necessary.

Practitioners and researchers must always rely on their own experience and
knowledge in evaluating and using any information, methods, compounds, or
experiments described herein. In using such information or methods they should be
mindful of their own safety and the safety of others, including parties for whom they
have a professional responsibility.

To the fullest extent of the law, neither the Publisher nor the authors, contributors, or
editors, assume any liability for any injury and/or damage to persons or property as a
matter of products liability, negligence or otherwise, or from any use or operation of
any methods, products, instructions, or ideas contained in the material herein.

ISBN: 978-0-12-804455-1

British Library Cataloguing-in-Publication Data
A catalogue record for this book is available from the British Library

Library of Congress Cataloging-in-Publication Data
A catalog record for this book is available from the Library of Congress

For information on all Academic Press publications
visit our website at www.elsevier.com

Typeset by MPS Limited, Chennai, India
www.adi-mps.com

Printed and bound in the United States

Publisher: Nikki Levy
Acquisition Editor: Nina Bandeira
Editorial Project Manager: Mariana Kuhl Leme
Editorial Project Manager Inter: Ana Claudia A. Garcia
Production Manager: Lisa Jones
Marketing Manager: Ofelia Chernock
Cover Designer: Victoria Pearson

To my family for steady support and understanding.

CONTENTS

PREFACE

Wild-growing mushrooms have been a part of my life since early childhood. A basket full of cepes and other valuable species was prestigious for us country boys. We passed through forests very early, even after daybreak. Our favorite dishes were only one of the rewards. The mood of the temperate forests with their calm, various aromas, and changeableness within a day and within seasons together with the esthetic look of mushrooms in their natural environment helped to form my stance in nature. Mushroom picking has remained my tried and true recreational activity, until now.

During my academic work as a food and feed chemist, mushrooms have become a part of my research. To tell the truth, this was never a funded project, it has only been a hobby. I have collected literature for decades, being focused on the chemical composition and nutritional value of both wild and cultivated edible mushrooms. And only in my senior age have I found the time and the courage to turn the expanding, but until now dispersed, information into a book. This book does not deal with medicinal and toxic species because data on these self-standing topics have already been collected.

Although written primarily for nutritionists and mushroom producers, it is my hope that this book will prove useful for students of food and human nutrition sciences and for mushroom fanciers.

Pavel Kalač
September 30, 2015
In České Budějovice

ACKNOWLEDGMENTS

I am particularly indebted to my colleagues Professor Martin Křížek, for his encouragement, Iveta Štefanová, MSc, for drawing chemical formulas and schemes, and Dr. Martin Šeda, for his help during communication with the editors. Mushroom images provided by Dr. Jan Borovička, Dr. Eva Dadáková, and Dr. Ivan Jablonský are acknowledged. Moreover, I highly appreciate the attitude and help of the Elsevier editors Ms. Nina Bandeira and Ms. Ana Claudia Abad Garcia.

BIOGRAPHY

Pavel Kalač is a Professor of Agricultural Chemistry at University of South Bohemia, České Budějovice, Czech Republic, where he has served on the faculty since 1971. He graduated from the Institute of Chemical Technology, Faculty of Food and Biochemical Technology, Prague. Professor Kalač has published 63 articles registered in Scopus (60 in Web of Science), including 36 articles and reviews in Elsevier journals (particularly on the topics of food chemistry (19) and meat science (8)).

He has published three books and numerous articles in Czech that deal with food and feed chemistry, and his work is frequently cited by researchers studying related topics.

The topic of edible mushroom chemistry has been his hobby for decades.

LIST OF FIGURES

LIST OF TABLES

CHAPTER 1

Introduction

Contents

Fresh and preserved mushrooms are consumed in many countries as a delicacy, particularly for their specific aroma and texture, but also for their low energy level and fiber content. However, mushrooms became a required part of nutrition during periods of staple food shortage, such as during wars. Approximately 14,000 mushroom species, described according to the rules of mycological nomenclature, represent approximately 10% of the estimated number of species existing on Earth. More than 2000 species are safe for consumption, and approximately 700 species are known to possess significant pharmacological properties (Wasser, 2002). Information on number of edible species collected for culinary purposes throughout the world varies widely between 200 and 3000. Approximately 100 species can be cultivated commercially, but only 10–20 of them can be cultivated on an industrial scale (Chang and Miles, 2004). The mushroom industry has three main segments: cultivated edible, wild-growing, and medicinal mushrooms.

According to FAOSTAT data (Table 1.1), the total world production of cultivated mushrooms was nearly 10 million tons in 2013, whereas it was only 4.2 million tons in 2000. China has been the leading producer by far. The most produced species is the *Agaricus bisporus* (white or button mushroom, brown mushroom, or portobello), dominating worldwide, followed by *Lentinula edodes* (commonly called by its Japanese name, shiitake), a species of genus

Edible Mushrooms.

1

Table 1.1 Production statistics of cultivated edible mushrooms, including truffles, in 2013 (FAOSTAT)

Country	Production (metric tons)	World production (%)
China	7,068,102	71.20
Italy	792,000	7.98
USA	406,198	4.09
Netherlands	323,000	3.25
Poland	220,000	2.22
Spain	149,700	1.51
France	104,621	1.05
Canada	81,788	0.82
United Kingdom	79,500	0.80
Ireland	93,600	0.64
Japan	61,500	0.62
World	9,926,966	100.00

Pleurotus (particularly *P. ostreatus*, oyster mushroom, hiratake), and *Flammulina velutipes* (golden needle mushroom, enokitake). Only approximately 45% of produced mushrooms are culinary-processed in the fresh form. The rest are preserved, mostly by canning and drying, with a ratio of approximately 10:1.

Consumption of wild-growing mushrooms has been preferred to cultivated species in many countries of Central and Eastern Europe due to species diversity and more savorous properties. Moreover, mushroom picking in forests and grasslands, as a lasting part of cultural heritage, has become a highly valued recreational activity in these countries. The collection of wild mushrooms can be seen as a relic of bygone gatherers and hunters. Moreover, mushrooms in their natural habitat are regarded for their esthetic value. This attitude is quite different from that of countries where wild mushrooms have been ignored as "toadstools."

Mushrooms have been collected mostly as a delicacy for the pickers' own consumption; however, the collection has been an economic activity for some rural populations. For instance, the picking is a "national hobby" in the Czech Republic. Interestingly, approximately 70% of the population picks mushrooms, with a statistical mean of 5–8 kg of fresh mushrooms per household or 2–3 kg per capita yearly. Some individuals consume more than 10 kg yearly. The factual consumption is lower due to removal of parts damaged by animals or insect larvae. In Finland, approximately 42% of households were engaged in picking, with a total harvest of 15,000 metric tons in 2011. Information on the harvest of wild-growing species worldwide is lacking.

Fresh mushrooms rank among the most perishable food materials, with a very short shelf life of only 1–3 days at ambient temperature. This considerably limits their distribution and marketing. Deterioration after harvesting, such as color changes, particularly browning, weight loss, texture changes, or cap opening, is caused by high water content, high respiration rate, and lack of physical protection to avoid water loss (transpiration) and microbial attack. Various endogenous enzymes participate in biochemical changes after disruption of cellular integrity such as by mechanical damage of tissues.

Quick deterioration of mushrooms has been an obstacle for both manufacturers and consumers. Drying, canning, or deep-freezing have been traditionally used for mushroom preservation; however, emerging technologies have been studied. Gamma irradiation has been shown to be a preservation method that saves chemical parameters of various mushroom species to a greater extent than drying or freezing (Fernandes et al., 2012). Nevertheless, the period between research results and industrial application can be rather long.

The following chapters provide recent overall knowledge and include data and references since 2000. It is not possible to cite all original works. Partial reviews, when available, are thus preferentially referred to and have numerous earlier references therein. Such reviews dealing with mushroom chemical composition and nutritional value were published during the past decade (Bernaś et al., 2006; Kalač, 2009, 2012, 2013; Wang et al., 2014). Most of the available data deal with European and East Asian edible species, often called culinary mushrooms.

Many mushroom species are toxic. The inability of immobile higher fungi to escape from an attack by fungivores, ranging from insects to mammals, has led to the evolution of several defense strategies to deter the pests. Fruit bodies of fungi often produce toxins and pungent or bitter compounds to deter fungivores (for an overview see Spiteller, 2008). This book does not focus on toxic mushroom species because such information is accessible elsewhere.

Similarly, this book does not deal with medicinal species, which are widely used in East Asian folk medicine and have been recently extensively studied as potential sources of novel drugs. Commercial products from medicinal mushrooms have been obtained from large-scale cultivation of fruit bodies. However, under such conditions it is difficult to sustain a regular level of effective compounds. Therefore, the production of fungal mycelium by submerged cultivation increases. The medicinal targets are very wide; antioxidant, antitumor, antidiabetic, antimalarial, antiviral, antimicrobial, anti-Alzheimer, and hypocholesterolemic activities have been observed. The specialized *International Journal of Medicinal Mushrooms* has been published since 1999. Several overviews of the topic are available (eg, De Silva et al., 2013; Lindequist et al., 2005; or Wasser, 2010).

1.1 BASIC MYCOLOGICAL TERMS

Scientific (Latin) names are used in the text because common names (eg, chanterelles, boletes, cepes, or truffles) are widely known for only the most consumed species, whereas less frequent mushroom species have more locally common names.

The term "mushroom" is used for a distinctive fruit body (mycocarp or carpophore) of a macrofungus (or higher fungus) that is large enough to be seen by the naked eye and to be picked up by hand. For culinary species, the term is used for the harvested, processed, and consumed part of macrofungi. Fruit

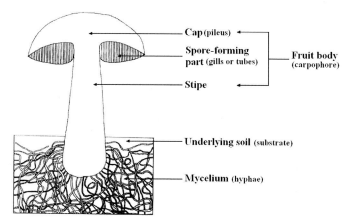

Figure 1.1 A sketch of a mushroom.

bodies are mostly above ground, varying in size, shape, and coloration for each mushroom species. The fruit body is the sexual part of a macrofungus bearing spores. It grows from spacious mycelia (hyphae), mostly underground, by the process of fructification. Mycelium is the vegetative part of a fungus, consisting of a mass of branched hyphae. The basic terminology of a fruit body is given in Fig. 1.1. The lifetime of the bulk of fruiting bodies is only approximately 10–14 days. In some species (eg, of genus *Comatus*) the life span is very short, even ephemeral.

Mushroom species can be divided into three classes according to their prevailing nutritional (trophic) strategy or ecological preferences. Ectomycorrhizal (symbiotic) species form a close, mutually profitable relationship with their host vascular plant, usually roots of a tree. Saprobic (saprotrophic) species (saprophytes) derive their nutrients from dead organic materials. Some species of this class are exploited for cultivation, whereas ectomycorrhizal species have not yet been successfully cultivated. The third group, parasitic species, lives on other species in a nonsymbiotic relationship, such as ligniperdous mushrooms on living trees.

REFERENCES

Bernaś, E., Jaworska, G., Lisiewska, Z., 2006. Edible mushrooms as a source of valuable nutritive constituents. Acta Sci. Pol., Technol. Aliment. 5, 5–20.

Chang, S.-T., Miles, P.G., 2004. Mushrooms: Cultivation, Nutritional Value, Medicinal Effect, and Environmental Impact, second ed. CRC Press, Boca Raton.

De Silva, D.D., Rapior, S., Sudarman, E., Stadler, M., Xu, J., Aisyah Alias, S., et al., 2013. Bioactive metabolites from macrofungi: ethnopharmacology, biological activities and chemistry. Fungal Divers. 62, 1–40.

Fernandes, Â., Antonio, A.L., Oliveira, M.B.P.P., Martins, A., Ferreira, I.C.F.R., 2012. Effect of gamma and electron beam irradiation on the physicochemical and nutritional properties of mushrooms: a review. Food Chem. 135, 641–650.

Kalač, P., 2009. Chemical composition and nutritional value of European species of wild growing mushrooms: a review. Food Chem. 113, 9–16.

Kalač, P. 2012. Chemical composition and nutritional value of European species of wild growing mushrooms. In Andres, S. & Baumann, N. (eds.). Mushrooms: Types, Properties and Nutrition. Nova Sci. Publ., New York, 129–152.

Kalač, P., 2013. A review of chemical composition and nutritional value of wild-growing and cultivated mushrooms. J. Sci. Food Agric. 93, 209–218.

Lindequist, U., Niedermeyer, T.H.J., Jülich, W.-D., 2005. The pharmacological potential of mushrooms. Evid. Based Complement. Alternat. Med. 2, 285–299.

Spiteller, P., 2008. Chemical defence strategies of higher fungi. Chem. Eur. J. 14, 9100–9110.

Wang, X.M., Yhang, J., Wu, L.H., Yhao, Z.L., Li, T., Li, J.Q., et al., 2014. A mini-review of chemical composition and nutritional value of edible wild-grown mushroom from China. Food Chem. 151, 279–285.

Wasser, S.P., 2002. Medicinal mushroom as a source of antitumor and immunomodulating polysaccharides. Appl. Microbiol. Biotechnol. 60, 258–274.

Wasser, S.P., 2010. Medicinal mushroom science: history, current status, future trends, and unsolved problems. Int. J. Med. Mushrooms 12, 1–16.

CHAPTER 2

Proximate Composition and Nutrients

Contents

Most data dealing with the nutritional composition of both cultivated and wild-growing mushrooms were published during the past decade. A great deal of the information in this chapter originates from numerous papers from the laboratory of Professor Isabel C.F.R. Ferreira, Polytechnic Institute of Bragança, Portugal.

2.1 DRY MATTER, PROXIMATE COMPOSITION, AND ENERGY VALUE

Overall data on the proximate (also crude or gross) composition and energy value of fresh wild-growing mushrooms are given in Table 2.1, and those for cultivated species are presented in Table 2.2. The data should be taken as approximate values for a few reasons: variability in mushroom chemical composition within the species is greater than in plants, and each individual fruit body

Table 2.1 Dry matter content (g 100 g⁻¹ fresh matter), proximate composition (g 100 g⁻¹ dry matter), and energy (kcal kg⁻¹ fresh matter) of unprocessed wild-growing mushrooms

Species	Dry matter	Crude protein	Crude fat	Ash	Carbohydrates	Energy	Reference
Agaricus albertii	9.3	19.8	1.4	22.1	56.7	318	Reis et al. (2014a)
Agaricus arvensis	5.1	56.3	2.7	3.5	37.5	204	Barros et al. (2007a)
Agaricus bitorquis	–	24.9	3.2	13.8	58.1	–	Glamočlija et al. (2015)
Agaricus campestris	14.9	38.9	2.7	3.5	54.9	562	Beluhan and Ranogajec (2011)
	11.8	18.6	0.1	23.2	58.1	364	Pereira et al. (2012)
	–	19.1	3.0	17.7	60.2	–	Glamočlija et al. (2015)
Agaricus macrosporus	–	21.9	2.4	10.4	65.3	–	Glamočlija et al. (2015)
Agaricus urinascens var. *excellens*	12.3	14.5	1.4	29.6	54.5	288	Reis et al. (2014a)
Amanita rubescens	9.1	26.0	7.2	4.6	62.2	380	Ouzouni and Riganakos (2007)
Amanita cesarea	–	6.3	6.4	14.8	72.5	–	Fernandes et al. (2015c)
Amanita crocea	11.0	20.0	4.6	25.7	49.7	351	Leal et al. (2013)
Amanita curtipes	–	6.4	8.6	17.2	67.8	–	Fernandes et al. (2015c)
Amanita mairei	23.2	17.7	8.3	11.2	62.8	920	Leal et al. (2013)
Amanita mairei	12.8	17.2	2.1	8.0	72.7	484	Ouzuni et al. (2009)
Armillariella mellea	11.7	16.4	5.6	6.8	71.2	514	Vaz et al. (2011)
	–	7.6	3.4	9.0	80.0	–	Ayaz et al. (2011a)
	–	17.3	6.6	12.3	63.8	–	Akata et al. (2012)

Boletus leucomelaena	–	16.0	2.2	6.3	75.5	–	Ayaz et al. (2011a)
Boletus aestivalis (syn. reticulatus)	8.9	22.6	2.6	19.7	55.1	297	Grangeia et al. (2011)
	–	27.9	3.1	16.6	52.4	–	Akata et al. (2012)
Boletus appendiculatus (syn. aereus)	12.4	19.1	4.5	6.3	70.1	–	Ouzuni et al. (2009)
	8.4	17.9	0.4	8.9	72.8	306	Heleno et al. (2011)
Boletus armeniacus	28.5	18.3	1.6	12.1	68.0	1053	Pereira et al. (2012)
Boletus edulis	12.0	26.5	2.8	5.3	65.4	471	Ouzouni and Riganakos (2007)
	12.2	36.9	2.9	5.3	54.9	436	Beluhan and Ranogajec (2011)
	–	22.8	2.9	5.0	69.3	–	Ayaz et al. (2011a)
	10.9	21.1	2.5	5.5	70.9	423	Heleno et al. (2011)
	10.0	22.6	3.8	5.2	68.4	377	Jaworska et al. (2012)
	9.0	16.4	5.0	8.0	70.6	375	Fernandes et al. (2014b)
Boletus erythropus	11.6	20.9	0.8	25.9	52.4	349	Grangeia et al. (2011)
Boletus fragrans	22.0	17.2	1.8	4.7	76.3	858	Grangeia et al. (2011)
Boletus regius	20.9	5.2	1.6	4.4	88.8	814	Leal et al. (2013)
Calocybe gambosa	13.9	36.7	1.3	8.0	54.0	467	Beluhan and Ranogajec (2011)
	9.1	15.5	0.8	13.9	69.8	318	Vaz et al. (2011)
Calvatia utriformis	22.0	20.4	1.9	17.8	59.9	744	Grangeia et al. (2011)

(Continued)

Table 2.1 Dry matter content (g 100 g⁻¹ fresh matter), proximate composition (g 100 g⁻¹ dry matter), and energy (kcal kg⁻¹ fresh matter) of unprocessed wild-growing mushrooms (Continued)

Species	Dry matter	Crude protein	Crude fat	Ash	Carbohydrates	Energy	Reference
Cantharellus cibarius	7.6	53.7	2.9	11.5	31.9	282	Barros et al. (2008a)
	17.4	15.1	2.9	9.4	72.6	656	Ouzuni et al. (2009)
	14.2	30.9	1.9	8.8	58.4	508	Beluhan and Ranogajec (2011)
Clitocybe odora	11.5	17.3	2.5	9.6	70.6	431	Vaz et al. (2011)
Coprinus comatus	14.8	15.7	1.1	12.9	70.3	525	Vaz et al. (2011)
	–	29.5	5.4	15.9	49.2	–	Akata et al. (2012)
	–	11.8	1.8	10.1	76.3	–	Stojković et al. (2013)
Craterellus cornucopioides	10.1	47.2	4.9	10.1	37.8	420	Beluhan and Ranogajec (2011)
Fistulina hepatica	12.8	21.0	5.9	8.4	64.7	507	Liu et al. (2012)
Flammulina velutipes	8.3	50.1	1.9	16.4	31.6	286	Heleno et al. (2009)
Gyromitra esculenta	9.3	17.9	1.8	9.4	70.9	346	Pereira et al. (2012)
Helvella lacunosa	14.3	14.7	0.7	32.1	52.5	394	Leal et al. (2013)
Hydnum repandum	17.6	4.4	2.4	21.7	71.5	573	Leal et al. (2013)
Laccaria amethystea	–	13.8	4.3	9.2	72.7	–	Ayaz et al. (2011a)
Laccaria laccata	13.9	29.8	2.8	10.3	57.1	518	Liu et al. (2012)
Lactarius deliciosus	11.8	62.8	3.8	20.7	12.7	397	Heleno et al. (2009)
	10.0	29.8	2.2	5.1	62.9	389	Barros et al. (2007a)
	10.0	17.9	6.5	14.3	61.3	371	Barros et al. (2007b)
	–	20.2	8.0	7.2	64.6	–	Akata et al. (2012)
	11.0	14.1	2.8	5.8	77.3	432	Kalogeropoulos et al. (2013)

Lactarius salmonicolor	12.3	37.3	2.0	23.3	37.4	389	Heleno et al. (2009)
	–	13.5	1.1	6.2	79.2	–	Akata et al. (2012)
Lactarius sanguifluus	9.1	17.0	3.1	7.3	72.6	352	Kalogeropoulos et al. (2013)
Lactarius semisanguifluus	10.4	15.2	3.7	5.7	75.4	412	Kalogeropoulos et al. (2013)
Laetiporus sulphureus	–	8.3	5.9	4.0	81.8	–	Ayaz et al. (2011a)
	–	16.0	2.4	9.0	72.6	–	Petrović et al. (2014)
	–	10.6	3.0	6.4	80.0	–	Kovács and Vetter (2015)
Lepista nuda	8.7	19.8	3.2	6.0	71.0	341	Ouzouni and Riganakos (2007)
	6.2	59.4	1.8	18.5	20.3	208	Barros et al. (2008a)
	8.7	24.1	3.2	6.0	66.7	341	Ouzuni et al. (2009)
Lycoperdon echinatum	14.8	23.5	1.2	9.4	65.9	544	Grangeia et al. (2011)
Lycoperdon perlatum	11.4	17.1	4.4	31.9	46.6	314	Barros et al. (2008a)
Macrolepiota mastoidea	11.3	21.9	2.6	8.0	67.5	381	Barros et al. (2007b)
Macrolepiota procera	12.3	23.9	2.3	5.4	68.4	480	Ouzouni and Riganakos (2007)
	10.0	7.6	1.5	9.9	81.0	365	Barros et al. (2007b)
	13.2	24.2	2.2	5.4	68.2	517	Beluhan and Ranogajec (2011)
	–	18.4	2.4	6.8	72.4	–	Ayaz et al. (2011a)
	14.1	19.0	2.9	8.0	70.1	540	Fernandes et al. (2013a)

(Continued)

Table 2.1 Dry matter content (g $100\,g^{-1}$ fresh matter), proximate composition (g $100\,g^{-1}$ dry matter), and energy (kcal kg^{-1} fresh matter) of unprocessed wild-growing mushrooms (Continued)

Species	Dry matter	Crude protein	Crude fat	Ash	Carbohydrates	Energy	Reference
Morchella esculenta	9.2	11.5	2.6	11.3	74.6	349	Heleno et al. (2013)
	10.6	11.5	2.3	7.9	78.3	403	Heleno et al. (2013)
Pleurotus eryngii	17.4	2.1	4.4	15.0	78.6	362	Reis et al. (2014a)
Pleurotus ostreatus	8.5	21.4	3.4	4.5	70.7	339	Ouzouni and Riganakos (2007)
	11.7	24.9	2.1	7.6	65.4	437	Beluhan and Ranogajec (2011)
Ramaria aurea	–	13.2	3.6	8.1	75.1	–	Akata et al. (2012)
	11.5	14.6	2.3	5.7	77.4	447	Pereira et al. (2012)
Ramaria botrytis	10.2	39.0	1.4	8.8	50.8	380	Barros et al. (2008a)
Russula aurea	20.0	10.3	1.2	12.8	75.7	710	Leal et al. (2013)
Russula cyanoxantha	15.6	16.8	1.5	7.0	74.7	590	Grangeia et al. (2011)
Russula delica	13.3	50.6	0.9	22.9	25.6	416	Heleno et al. (2009)
	10.3	15.2	1.3	8.3	75.2	385	Kalogeropoulos et al. (2013)
Russula olivacea	8.0	13.8	3.4	8.8	74.0	363	Fernandes et al. (2014b)
	15.4	16.8	2.0	37.8	43.4	399	Grangeia et al. (2011)
Russula virescens	7.5	21.9	1.9	11.0	62.2	274	Leal et al. (2013)
Suillus bellinii	5.1	17.2	3.9	9.0	69.9	196	Kalogeropoulos et al. (2013)

Species							Reference
Suillus granulatus	7.7	16.5	4.0	5.2	74.3	307	Ouzouni and Riganakos (2007)
from Portugal	–	14.8	3.7	8.0	73.5	–	Reis et al. (2014b)
from Serbia	–	7.9	0.3	10.4	81.4	–	Reis et al. (2014b)
Suillus imbricatus	6.1	30.0	3.5	12.1	54.4	369	Barros et al. (2007b)
Suillus luteus	4.5	25.1	7.0	18.3	49.6	164	Jaworska et al. (2014)
Suillus variegatus	9.2	17.6	3.3	15.4	63.7	328	Pereira et al. (2012)
Terfezia boudieri	14.4	15.0	9.9	11.5	63.6	565	Dundar et al. (2012)
Termitomyces robustus	–	13.3	1.3	6.3	79.1	–	Obodai et al. (2014)
Tricholoma imbricatum	17.6	50.5	1.9	6.5	41.1	677	Heleno et al. (2009)
Tricholoma portentosum	–	13.7	5.8	9.9	70.6	–	Díez and Alvarez (2001)
Tricholoma terreum	7.0	30.5	5.5	11.7	52.3	265	Barros et al. (2007a)
	–	14.1	6.6	12.1	67.2	–	Díez and Alvarez (2001)
Tuber aestivum	–	19.1	2.3	7.6	71.0	–	Kruzselyi and Vetter (2014)

Contents of crude protein were recalculated for data from some works using a factor of 6.25. A factor of 4.38 is thought to be credible for mushrooms. The content of carbohydrates was calculated as: [100 – (crude protein + fat + ash)]. Energy (kcal) was calculated as: 4 (g protein + g carbohydrates) + 9 (g fat). 1 kcal = 4.168 kJ.

Table 2.2 Dry matter content (g 100 g^{-1} fresh matter), proximate composition (g 100 g^{-1} dry matter), and energy (kcal kg^{-1} fresh matter) of unprocessed cultivated mushrooms

Species	Dry matter	Crude protein	Crude fat	Ash	Carbohydrates	Energy	Reference
Agaricus bisporus							
white	7.7	27.1	4.3	10.1	58.5	293	Mattila et al. (2002)
	8.7	14.1	2.2	9.7	74.0	325	Reis et al. (2012)
	8.6	13.8	4.0	10.7	71.5	377	Jaworska et al. (2015)
brown	7.8	26.5	4.0	10.0	59.5	296	Mattila et al. (2002)
	8.4	15.4	1.7	11.4	71.5	303	Reis et al. (2012)
unspecified	–	36.3	0.8	12.0	50.9	–	Akyüz and Kirbağ (2010)
	–	26.5	2.5	8.8	62.2	–	Ulziijargal and Mau (2011)
	8.8	30.7	4.0	8.1	57.2	309	Jaworska et al. (2012)
	7.3	26.3	3.2	7.2	63.3	387	Liu et al. (2014)
	–	3.1	1.8	6.0	89.1	–	Stojković et al. (2014)
	–	10.0	3.1	15.0	71.9	–	Glamočlija et al. (2015)
Agaricus subrufescens	–	26.7	2.6	6.8	63.9	–	Tsai et al. (2008)
	–	31.3	1.8	7.5	59.4	–	Carneiro et al. (2013)
	–	28.0	2.6	9.5	59.9	–	Cohen et al. (2014)
	–	13.4	2.8	8.2	75.6	–	Stojković et al. (2014)
Auricularia auricula-judae	–	8.1	1.5	9.4	81.0	–	Cheung (2013)
	–	19.3	0.8	3.1	76.8	–	Obodai et al. (2014)
Coprinus comatus	–	11.0	2.0	10.5	76.5	–	Stojković et al. (2013)
	–	16.6	1.4	13.8	68.2	–	Cohen et al. (2014)

Flammulina velutipes							
white	10.9	20.0	8.9	6.9	64.2	454	Yang et al. (2001)
yellow	12.8	26.7	9.2	7.5	56.6	532	Yang et al. (2001)
	–	26.7	9.2	7.5	56.6	–	Ulzijargal and Mau (2011)
Grifola frondosa	12.1	3.9	2.9	7.3	85.9	467	Reis et al. (2012)
	–	17.2	2.2	8.7	71.9	–	Cohen et al. (2014)
Hericium erinaceus	–	13.4	5.6	4.9	76.1	–	Cohen et al. (2014)
Hypsizygus marmoreus	–	15.5	5.4	7.2	71.9	–	Cohen et al. (2014)
	–	19.6	4.1	7.8	68.5	–	Lee et al. (2009)
Lentinula edodes	18.2	20.5	6.3	5.3	67.9	747	Yang et al. (2001)
	8.4	21.4	3.7	5.8	69.1	332	Mattila et al. (2002)
	–	20.5	6.3	5.3	67.9	–	Ulzijargal and Mau (2011)
	20.2	4.4	1.7	6.7	87.2	772	Reis et al. (2012)
	–	12.8	1.0	4.3	81.9	–	Carneiro et al. (2013)
	–	14.0	0.9	5.5	79.6	–	Cohen et al. (2014)
Lentinula squarrosulus	–	16.0	1.9	5.9	76.2	–	Obodai et al. (2014)
Oudemansiella submucida	–	27.0	3.4	5.8	63.8	–	Zhou et al. (2015)
	–	15.3	7.4	11.9	65.4	–	Zhou et al. (2015)
Pleurotus cystidiosus	13.3	15.4	3.1	9.6	71.9	502	Yang et al. (2001)

(Continued)

Table 2.2 Dry matter content (g 100 g⁻¹ fresh matter), proximate composition (g 100 g⁻¹ dry matter), and energy (kcal kg⁻¹ fresh matter) of unprocessed cultivated mushrooms (Continued)

Species	Dry matter	Crude protein	Crude fat	Ash	Carbohydrates	Energy	Reference
Pleurotus ostreatus	9.7	27.5	–	8.3	–	–	Manzi et al. (1999)
	11.4	23.9	2.2	7.6	66.3	434	Yang et al. (2001)
	8.0	24.6	4.4	8.0	63.0	312	Mattila et al. (2002)
	–	41.6	0.5	6.0	51.9	–	Akyüz and Kirbağ (2010)
		23.9	2.2	7.6	66.3	–	Ulziijargal and Mau (2011)
	8.8	16.7	5.5	6.7	71.1	433	Jaworska et al. (2011)
	10.8	7.0	1.4	5.7	85.9	416	Reis et al. (2012)
	–	25.6	2.5	7.7	64.2	–	Cohen et al. (2014)
	–	28.4	1.5	6.0	64.1	–	Obodai et al. (2014)
	15.7	14.7	1.5	5.7	78.1	604	Fernandes et al. (2015a)
	10.0	8.3	8.0	8.2	75.5	407	Jaworska et al. (2015)
Pleurotus eryngii	8.3	22.9	–	9.9	–	–	Manzi et al. (1999)
	–	22.2	1.6	5.8	70.4	–	Ulziijargal and Mau (2011)
	11.0	11.0	1.5	6.2	81.3	421	Reis et al. (2012)
	6.9	19.1	2.6	6.8	71.5	266	Cui et al. (2014)
	11.2	18.2	2.5	6.7	72.6	386	Li X. et al. (2015)
Pleurotus sajor-caju	–	37.4	1.0	6.3	55.3	–	Akyüz and Kirbağ (2010)
	–	29.3	0.9	6.8	63.0	–	Gogavekar et al. (2014)
	–	15.3	2.9	6.4	75.4	–	Obodai et al. (2014)
Pleurotus tuber-regium	–	13.3	1.3	6.3	79.1	–	Obodai et al. (2014)
Tremella aurantialba	–	7.8	3.5	6.1	82.6	–	Zhou et al. (2015)

Contents of crude protein were recalculated for data from works using a factor of 6.25. A factor of 4.38 is thought to be credible for mushrooms. The content of carbohydrates was calculated as: [100 − (crude protein + fat + ash)]. Energy (kcal) was calculated as: 4 (g protein + g carbohydrates) + 9 (g fat). 1 kcal = 4.168 kJ.

can result from the cross-breeding of different hyphaes, thus presenting a distinct genotype.

The content of crude protein has been quantified by the Kjeldahl method determining total nitrogen. In nutritionally high-value proteins, such as in meats, nitrogen comprises 16% of protein weight. A factor of 6.25 (ie, 100:16) has therefore been used for the calculation of protein content from the level of determined nitrogen. However, mushrooms differ from such foods by a high proportion of nonprotein nitrogen, particularly nitrogen-containing polysaccharide chitin. A lower factor of 4.38 has therefore been recommended for the calculation of mushroom protein content. This factor was used for all protein values given in Tables 2.1 and 2.2. Thus, the previous data regarding mushroom protein values were overestimated by nearly one-third.

Total carbohydrate contents in dry matter have not been determined but are calculated as [100 − (crude protein + crude fat + ash)]. The calculated content comprises various carbohydrates, including indigestible ones such as dietary fiber. The equation [kcal = 4 × (g crude protein + g carbohydrates) + 9 × (g crude fat)] commonly used in the references cited in Tables 2.1 and 2.2 for the calculation thus overestimates the total energy level. Therefore, the reported values should be reduced by up to one-third (see Section 2.4.2).

According to the data of Tables 2.1 and 2.2, considerable differences in the composition are evident not only among species but also within a species if comparing data of various laboratories. The differences can be partly caused by the analysis of fruit bodies at different stages of development. Nevertheless, such data have been limited. For instance, Cui et al. (2014) reported only a small increase of protein content and decrease of fat proportion during five stages, from bud-forming to full maturity, of *Pleurotus eryngii*.

Dry matter (DM; or dry weight) of both wild-growing and cultivated mushrooms is very low, commonly ranging between 8 and 14 g $100\,g^{-1}$ fresh matter (FM; or fresh weight). The value of $10\,g\ 100\,g^{-1}$ FM (10%) is widely used for the calculations if actual DM content is unknown. High water content and water activity participate considerably in the short shelf life of fruit bodies. Dried mushrooms rank among food materials with high hygroscopicity.

The most frequent values for crude protein, crude fat, and ash (minerals) contents are within the ranges of 20–25, 2–3, and 5–12 $g\ 100\,g^{-1}$ DM, respectively, with carbohydrates forming the rest of the DM. Ash content is the steadiest value. Results from the comparison of data for wild and cultivated mushrooms indicate that the differences both among and within species seem to be more important than the effect of provenance. In cultivated species, the effect of the used substrate is of significance in both yield and composition.

Low DM and fat content result in low energy of approximately 350–450 $kcal\,kg^{-1}$ FM, with great variations given chiefly by varying DM content. Such calculated values have to be reduced by up to one-third. Overall, both wild and cultivated mushrooms are a low-energy delicacy.

The very scarce information on mushroom digestibility and bioavailability of their components has been a weak point in the evaluation of the nutritional value. It may be supposed that a high proportion of indigestible dietary fiber apparently limits availability of other components. Moreover, most data are given for fresh mushrooms, whereas information on the nutrient changes under various storing and cooking conditions remains very limited. It seems that boiled mushrooms have a lower energy value than dried or frozen mushrooms.

Overall, despite rapidly growing quantitative compositional data, it is not yet possible to draw generalized conclusions about the relations between mushroom proximate chemical composition and their nutritional value.

Synoptic information on composition and biological properties of truffles, which are apart of common culinary species due to their extreme price, is available (Wang and Marcone, 2011).

2.2 PROTEINS

The information making the public familiar with the nutritional value of mushrooms underlines their high protein content. Mushrooms were once called "meat of the poverty" in some countries of Central Europe. However, such an opinion has to be corrected.

As mentioned, the levels of crude protein given in Tables 2.1 and 2.2 were calculated using a factor of 4.38 for the multiplication of total nitrogen content determined by Kjeldahl method. Nevertheless, protein content obtained in this manner is unreliable because it is not based on strict analytical data but rather on common agreement. The nutritional value can be overestimated. Braaksma and Schaap (1996) reported for *Agaricus bisporus* no fewer than four-times lower protein content, as determined by another analytical method, than that resulting from Kjeldahl total nitrogen. Bauer Petrovska (2001) concluded a somewhat lower factor of 4.16 for 52 Macedonian species, mostly wild-growing, with a mean proportion of 33.4% nonprotein nitrogen from total nitrogen.

Data on crude protein contents in Tables 2.1 and 2.2 vary considerably within species, such as for *Cantharellus cibarius*, *Lepista nuda*, or *A. bisporus*. The variability is even more apparent from values listed in Table 2.3. Unfortunately, the factors affecting the variability have remained poorly explained until now. Vetter and Rimóczi (1993) reported the highest crude protein content together with the highest digestibility of 92% in cultivated *Pleurotus ostreatus* (oyster mushroom) with a cap diameter of 5–8 cm. At that stage of development, crude protein contents were 36.4% and 11.8% in caps and stipes, respectively. Thereafter, both crude protein and its digestibility decreased. Thus, further studies on the changes in bioavailable nutritional protein content during the development of both cultivated and wild-growing mushrooms are needed.

Information on protein fractions has been very scarce. Bauer Petrovska (2001) reported mean proportions of 24.8%, 11.5%,

Table 2.3 Variability of crude protein content in several mushroom species

Species	Number of samples	Mean (% of dry matter)	Standard deviation (% of dry matter)	Relative standard deviation (%)
Armillariella mellea	6	22.3	5.4	24.2
Boletus edulis	10	33.1	3.1	9.4
Cantharellus cibarius	8	18.7	5.9	31.5
Clitocybe nebularis	6	39.0	4.4	11.3
Craterellus cornucopioides	6	22.3	5.5	24.7
Leccinum scabrum	9	30.5	4.1	13.4
Marasmius oreades	3	52.8	2.1	4.0
Xerocomus subtomentosus	7	33.2	6.8	20.5

Adapted from Vetter (1993a). With permission.

7.4%, 11.5%, 5.7%, 5.3%, and 33.8% of total protein for albumins, globulins, glutelin-like matter, glutelins, prolamins, prolamine-like matter, and residues, respectively, in 24 mushroom species. The respective values (except for residues) were 57.8%, 8.6%, 2.1%, 25.7%, 3.2%, and 2.5% in valued *Tuber aestivum* (Kruzselyi and Vetter, 2014) and 51.3%, 11.1%, 13.2%, 17.1%, 4.4%, and 2.9% in wood-decaying *Laetiporus sulphureus* (formerly *Polyporus sulphureus*) (Kovács and Vetter, 2015). Also, Florczak et al. (2004) observed albumins as the prevailing protein fraction in *Armillariella mellea*, *Coprinus atramentarius*, and *Tricholoma equestre*.

Table 2.4 and the data of Ayaz et al. (2011b) indicate that amino acid composition varies among mushroom species. However, information on variability within a species and on factors affecting the proportion of various proteins would be useful. The proportion of essential (indispensable) amino acids varies around 40% from total amino acid content in wild species, whereas it varies between 30% and 35% in cultivated mushrooms. Methionine appears to be the very limiting essential amino acid. Levels of histidine and arginine, called semi-essential amino acids because they are necessary in the diet during childhood, vary widely. Aspartic acid and glutamic acid occur in relatively high

Table 2.4 Proportion of essential amino acids (% of total amino acids) in mushroom proteins

Species	Val	Leu	Ile	Thr	Met	Lys	Phe	Trp	Total	Reference
Wild-growing species										
Agaricus arvensis	6.4	8.1	4.9	4.7	1.5	6.3	5.6	–	37.5	Vetter (1993b)
Agaricus silvaticus	3.8	6.9	4.5	4.5	1.2	6.0	3.8	–	30.7	Vetter (1993b)
Cantharellus cibarius	3.5	16.3	3.3	4.2	1.0	4.3	3.2	1.7	35.8	Surinrut et al. (1987)
Hydnum repandum	3.9	14.5	3.2	4.4	1.0	4.2	3.4	1.4	34.5	Surinrut et al. (1987)
Russula cyanoxantha	6.8	7.8	4.9	5.0	2.0	6.3	7.8	–	40.6	Vetter (1993b)
Russula vesca	7.1	8.4	5.5	5.0	0.9	7.0	6.2	–	40.1	Vetter (1993b)
Tricholoma portentosum	7.8	9.4	3.7	9.5	3.0	8.6	4.4	1.0	46.4	Díez and Alvarez (2001)
Tricholoma terreum	8.9	8.2	3.6	9.1	3.5	7.6	6.6	1.1	47.5	Díez and Alvarez (2001)
Cultivated species										
Agaricus bisporus	3.9	6.0	3.1	3.4	1.4	6.1	3.1	–	27.0	Vetter (1993b)
white	5.0	6.3	3.8	4.6	1.4	5.9	4.4	–	31.4	Mattila et al. (2002)
brown	5.0	6.1	3.6	4.3	1.3	5.3	4.9	–	30.5	Mattila et al. (2002)
Lentinula edodes	3.8	6.4	3.3	5.6	2.2	5.0	3.8	1.9	32.0	Manzi et al. (1999)
Pleurotus eryngii	3.9	7.1	3.7	5.3	1.7	6.8	4.2	1.4	34.1	Manzi et al. (1999)
	6.0	6.3	3.8	4.7	1.4	5.9	4.4	–	32.5	Mattila et al. (2002)
Pleurotus ostreatus	4.7	6.8	4.3	5.0	1.9	6.0	4.3	1.4	34.4	Manzi et al. (1999)
	4.9	6.1	3.6	4.6	1.5	5.5	4.9	–	31.1	Mattila et al. (2002)
Standard protein (FAO/WHO)	5.0	7.0	4.0	4.0	3.5	5.4	6.1	1.0	36.0	

Val, valine; Leu, leucine; Ile, isoleucine; Thr, threonine; Met, methionine; Lys, lysine; Phe, phenylalanine; Trp, tryptophan.

proportion within the nonessential amino acids. Overall, data suggest a higher nutritional value of mushroom proteins as compared with most plant proteins, but it is lower than that in high-value proteins of hen egg whites, milk, or meat.

Mushrooms also contain a low level of free amino acids (ie, unbound in proteins). These amino acids contribute to the taste of mushrooms and are discussed in Section 3.1.1.

Data regarding changes of protein and amino acid composition during mushroom preservation and cooking have been limited until now. In a series of experiments with *A. bisporus*, *Boletus edulis*, and *P. ostreatus*, Jaworska et al. (2011, 2012) and Jaworska and Bernaś (2012) tested the effect of pretreatments (blanching or soaking and blanching) followed by either freezing or canning and 12-month storage on amino acid composition. Both increases and decreases of the individual amino acids were observed, particularly among nonessential ones. The nutritional value of preserved mushrooms was thus only somewhat lower than that of raw materials.

Mushrooms have recently become a promising source of novel proteins with unique features, particularly for medicine and biotechnology. Such proteins are briefly mentioned in Section 4.4.

2.3 LIPIDS

Mushroom lipids are formed by two groups, neutral and polar lipids. Neutral lipids comprise fats, esters of trifunctional alcohol glycerol, and fatty acids, with prevailing fully esterified glycerol (triglycerides or triacylglycerols) and waxes, esters of a monofunctional higher alcohol, and a fatty acid. Phospholipids, the main polar lipids occurring in mushrooms, are, in their chemical structure, related to fats, but instead of one bound fatty acid they contain, as an ester, bound phosphoric acid esterified with an aminoalcohol, particularly ethanolamine or choline.

Information on the proportion of neutral and polar lipids has been scarce. Pedneault et al. (2008) reported the mean proportion of neutral lipids as 34.4% from total lipids for 10 Canadian wild species. However, there is a wide variation ranging between 13.2% in *Agaricus campestris* and 67.0% in *Amanita rubescens*. Data on waxes have been virtually lacking.

More than 40 fatty acids were observed in various mushroom species. The prevailing fatty acids are aliphatic and monocarboxylic, with a straight, even carbon chain; unsaturated acids are of *cis*-configuration (see Appendix III). Fatty acids with an odd number of carbons with branched chain or hydroxy fatty acids are only minor components. Nevertheless, they are regarded as a potential marker for distinguishing the individual species and are useful for both taxonomy and for the identification of adulteration (Barreira et al., 2012). Elaidic acid, a *trans*-isomer of oleic acid, was first reported in mushrooms by Pedneault et al. (2008) at a proportion of 0.07–0.48% from total fatty acids.

Fatty acid composition of lipids in many mushroom species, particularly wild-growing, has been determined during the past years. Selected data on seven nutritionally interesting acids and their groups are presented in Table 2.5. Data from different laboratories available for several species often show considerable differences, such as for *A. rubescens* or *Suillus granulatus*. Environmental factors affecting changes in fatty acid composition within a species remain unexplained. Genus *Agaricus* spp. seem to be lower in oleic acid than in most other species. Unsaturated fatty acids prevail, formed mostly by alpha-linoleic acid from the ω-6 family and by oleic acid. However, nutritionally desirable linolenic acid from the ω-3 family occurs at very low levels. The ω-3:ω-6 ratio of fatty acids is thus far from the nutritional optimum of 1:<5. An extreme proportion of linolenic acid in cultivated *Flammulina velutipes* was not proven in its wild-grown counterpart. Among saturated fatty acids, which are nutritionally unwanted, there prevails palmitic acid, whereas the most undesirable myristic acid occurs at only very low levels.

Linoleic acid is a precursor of numerous oxidation products with the characteristic attractive smell of mushrooms, particularly

Table 2.5 Proportion of major fatty acids (% of total fatty acids) in selected mushroom species

Species	Lauric acid	Myristic acid	Palmitic acid	Stearic acid	Oleic acid	Linoleic acid	Linolenic acid	SFA	MUFA	PUFA	Reference
Wild-growing species											
Agaricus albertii	–	–	11.1	3.1	2.1	75.8	–	21.1	2.4	76.5	Reis et al. (2014a)
Agaricus arvensis	–	2.3	14.6	3.4	15.5	56.1	0.2	23.5	19.9	56.6	Barros et al. (2007a)
	0.1	0.8	20.2	5.5	4.5	58.7	0.1	33.4	7.3	59.3	Pedneault et al. (2008)
Agaricus bitorquis	<0.1	0.7	12.7	5.0	5.5	69.9	0.9	23.0	6.0	71.0	Glamočlija et al. (2015)
Agaricus campestris	0.1	0.5	16.5	3.0	4.0	68.8	0.2	24.5	6.1	69.4	Pedneault et al. (2008)
	–	–	12.5	2.7	6.1	69.0	–	20.9	9.1	70.0	Pereira et al. (2012)
	0.1	0.8	13.2	3.5	3.5	71.4	0.2	22.6	5.3	72.1	Glamočlija et al. (2015)
Agaricus macrosporus	Traces	0.3	10.9	3.1	2.6	74.9	0.2	20.8	2.9	76.3	Glamočlija et al. (2015)
Agaricus silvaticus	Traces	0.3	11.7	1.4	6.7	74.8	0.1	17.1	7.7	75.2	Barros et al. (2008b)
Agaricus silvicola	Traces	0.3	10.0	2.6	3.5	76.5	<0.1	18.8	4.3	76.9	Barros et al. (2008b)
Agaricus urinascens var. excellens	–	–	14.9	3.6	5.5	51.2	–	28.8	19.2	52.0	Reis et al. (2014a)
Amanita caesarea	Traces	0.2	15.0	6.1	58.0	19.0	Traces	21.5	59.1	19.4	Doğan and Akbaş (2013)
	–	0.3	14.5	2.9	44.0	35.0	–	19.4	45.6	35.0	Fernandes et al. (2015c)

Species											Reference
Amanita ceciliae	Traces	0.2	16.4	3.6	42.7	31.8	Traces	22.0	44.1	33.9	Akata et al. (2013)
Amanita curtipes	–	0.3	18.9	4.1	54.0	19.2	–	25.4	55.2	19.4	Fernandes et al. (2015c)
Amanita ovoidea	<0.1	0.2	15.8	5.4	58.8	18.0	<0.1	21.8	59.4	18.8	Doğan (2013)
Amanita rubescens	0.1	0.2	14.5	4.5	58.0	19.0	0.1	21.8	59.0	19.2	Pedneault et al. (2008)
	Traces	0.8	22.7	16.1	40.4	21.1	ND	46.8	41.7	11.5	Ribeiro et al. (2009)
Armillariella mellea	0.7	0.3	11.0	3.5	47.7	27.7	<0.1	17.2	55.0	27.8	Vaz et al. (2011)
	0.6	0.4	12.6	3.3	22.7	48.6	ND	23.8	26.5	49.7	Akata et al. (2013)
Armillariella polytricha	–	0.8	11.2	10.2	27.1	29.5	3.6	39.8	27.1	33.1	Kavishree et al. (2008)
Boletus aestivalis	<0.1	0.4	11.3	5.8	37.6	36.6	ND	21.1	40.3	38.6	Ergönül et al. (2013)
Boletus	<0.1	0.15	12.5	3.8	36.7	43.8	Traces	18.1	37.8	44.1	Heleno et al. (2011)
Boletus appendiculatus	–	–	15.7	2.9	27.6	49.0	–	21.0	29.7	49.3	Pereira et al. (2012)
Boletus armeniacus	0.1	0.2	9.8	2.7	36.1	42.2	0.2	15.5	41.4	43.1	Pedneault et al. (2006)
Boletus edulis	Traces	0.15	10.0	2.75	39.7	44.3	<0.1	14.5	40.9	44.6	Barros et al. (2008b)
	–	ND	21.6	9.1	31.1	33.8	3.6	33.4	31.1	35.5	Kavishree et al. (2008)
	–	–	8.9	3.0	29.4	51.7	1.2	–	–	–	Dembitsky et al. (2010)
	Traces	0.1	9.6	3.1	42.1	41.3	<0.1	14.8	43.5	41.7	Heleno et al. (2011)
	–	0.35	11.0	0.92	5.7	77.2	1.76	13.2	7.4	79.3	Fernandes et al. (2014b)

(Continued)

Table 2.5 Proportion of major fatty acids (% of total fatty acids) in selected mushroom species (Continued)

Species	Lauric acid	Myristic acid	Palmitic acid	Stearic acid	Oleic acid	Linoleic acid	Linolenic acid	SFA	MUFA	PUFA	Reference
Boletus erythropus	0.6	1.4	21.3	4.2	14.7	48.8	1.1	33.6	16.1	50.3	Grangeia et al. (2011)
Boletus regius	–	–	15.9	1.6	21.8	56.1	–	19.5	24.0	56.5	Leal et al. (2013)
Calocybe gambosa	0.2	0.4	15.2	2.1	18.1	57.8	0.5	22.5	19.1	58.4	Barros et al. (2008b)
	0.2	0.4	13.6	3.2	32.5	43.9	0.9	21.5	33.4	45.1	Vaz et al. (2011)
Calvatia utriformis	0.2	0.5	13.5	2.4	6.0	70.3	0.6	22.4	6.5	71.1	Grangeia et al. (2011)
Cantharellus cibarius	Traces	0.1	7.2	3.3	8.1	50.0	0.1	12.0	37.5	50.5	Barros et al. (2008a)
	Traces	0.1	13.1	6.5	10.8	53.6	0.1	22.6	23.3	54.1	Barros et al. (2008b)
	–	8.0	18.3	6.0	35.4	17.3	ND	39.0	43.7	17.3	Kavishree et al. (2008)
	Traces	0.3	1.9	0.8	17.8	78.8	ND	3.2	17.9	78.9	Ribeiro et al. (2009)
	0.1	0.5	17.1	5.6	21.1	49.6	Traces	26.0	22.9	51.1	Akata et al. (2013)
Clitocybe odora	0.1	0.2	12.5	3.5	46.1	34.9	0.9	18.6	46.4	35.0	Vaz et al. (2011)
Coprinus comatus	0.1	0.6	18.9	1.8	7.5	64.5	0.5	23.8	10.4	65.2	Pedneault et al. (2008)
	0.2	0.4	10.6	1.9	6.3	74.9	1.9	15.4	7.1	77.5	Vaz et al. (2011)
	–	–	12.9	2.3	13.5	64.1	–	18.7	15.3	66.0	Stojković et al. (2013)
Craterellus cornucopioides	<0.1	<0.1	6.7	7.8	51.9	23.7	<0.1	16.4	59.9	23.7	Barros et al. (2008b)
	–	–	10.2	5.4	54.0	28.4	1.3	15.8	54.4	29.8	Liu et al. (2012)
Flammulina velutipes	–	–	10.3	1.4	15.1	56.3	–	14.4	17.6	68.0	Pereira et al. (2012)
	0.3	0.5	14.6	3.6	16.4	40.9	Traces	20.7	18.6	60.7	Ergönül et al. (2013)

Species											Reference
Helvella crispa	–	ND	10.6	ND	22.7	62.7	0.3	12.8	24.2	63.0	Kavishree et al. (2008)
Hericium coralloides	–	–	23.3	6.8	33.7	30.9	–	33.5	34.6	31.9	Heleno et al. (2015)
Hericium erinaceus	–	–	37.6	7.6	26.1	25.1	–	47.6	26.8	25.6	Heleno et al. (2015)
Hydnum repandum	–	1.6	15.7	0.9	26.4	27.2	20.3	24.6	27.9	47.5	Kavishree et al. (2008)
Laccaria amethystea	–	–	6.9	3.3	13.7	74.4	1.1	10.3	13.7	76.0	Liu et al. (2012)
Laccaria laccata	<0.1	0.1	11.6	2.0	60.7	20.5	0.4	15.4	63.6	21.0	Heleno et al. (2009)
Lactarius deliciosus	–	0.5	12.1	25.3	41.3	17.1	0.3	40.1	42.3	17.6	Barros et al. (2007a)
	–	2.0	16.3	6.1	33.0	37.1	ND	28.5	34.4	37.1	Kavishree et al. (2008)
Lactarius sanguifluus	0.3	0.2	9.1	29.8	21.8	31.8	0.6	42.6	22.3	35.1	Kalogeropoulos et al. (2013)
	–	–	9.7	41.4	25.9	22.9	–	51.2	25.9	22.9	Öztürk et al. (2014)
	–	0.3	23.1	4.9	32.4	35.1	ND	31.2	33.7	35.1	Kavishree et al. (2008)
Laetiporus sulphureus	0.2	0.1	7.1	34.4	18.8	34.6	0.7	44.0	19.2	36.8	Kalogeropoulos et al. (2013)
	0.1	0.5	11.7	3.2	14.5	63.3	0.2	20.5	15.3	64.2	Petrović et al. (2014)
Leccinum aurantiacum	0.1	0.7	16.5	0.9	21.1	52.9	0.2	22.2	24.1	53.7	Pedneault et al. (2006)

(Continued)

Table 2.5 Proportion of major fatty acids (% of total fatty acids) in selected mushroom species (Continued)

Species	Lauric acid	Myristic acid	Palmitic acid	Stearic acid	Oleic acid	Linoleic acid	Linolenic acid	SFA	MUFA	PUFA	Reference
Leccinum scabrum	Traces	0.5	13.8	1.1	37.8	42.3	0.1	17.5	39.4	43.1	Pedneault et al. (2006)
	–	–	9.7	3.4	31.7	45.8	3.1	–	–	–	Dembitsky et al. (2010)
Lepista nuda	0.1	0.3	11.8	2.4	29.5	51.5	0.2	17.6	30.3	52.1	Barros et al. (2008a)
Leucopaxillus giganteus	–	2.7	13.5	2.1	21.1	46.2	0.1	19.3	34.1	46.6	Barros et al. (2007a)
Lycoperdon perlatum	0.2	0.4	12.9	3.0	4.6	70.7	0.2	23.6	4.9	71.5	Barros et al. (2008a)
Macrolepiota procera	–	2.8	4.6	ND	17.2	47.0	15.6	20.2	17.2	62.6	Kavishree et al. (2008)
	–	0.3	21.0	1.6	7.0	65.0	–	25.0	9.0	66.0	Fernandes et al. (2014a)
Marasmius oreades	<0.1	0.15	13.8	1.7	51.9	23.7	<0.1	18.9	30.2	50.9	Barros et al. (2008b)
Morchella conica	–	ND	8.5	5.4	11.3	68.6	ND	18.5	12.9	68.6	Kavishree et al. (2008)
Morchella esculenta	–	–	9.5	2.6	12.4	71.8	0.2	13.7	13.8	72.5	Heleno et al. (2013)
Pleurotus djamor	–	1.6	15.8	ND	28.8	45.5	ND	25.1	29.4	45.5	Kavishree et al. (2008)
Pleurotus eryngii	–	–	17.4	4.8	47.5	24.7	–	25.8	49.0	25.2	Reis et al. (2014a)
Pleurotus ostreatus	0.2	0.7	12.4	3.7	10.4	65.3	Traces	21.8	11.4	66.8	Ergönül et al. (2013)
Pleurotus sajor-caju	–	0.6	15.8	ND	16.4	53.8	ND	27.1	19.1	53.8	Kavishree et al. (2008)

Species											Reference
Ramaria flava	2.2	3.6	14.4	6.7	47.1	9.8	–	39.1	47.9	13.0	Öztürk et al. (2014)
Ramaria botrytis	Traces	0.4	9.9	2.4	43.9	38.3	Traces	16.4	44.7	38.9	Barros et al. (2008a)
Russula cyanoxantha	0.4	0.4	13.0	11.1	28.4	43.7	0.1	26.9	29.1	44.0	Grangeia et al. (2011)
Russula delica	–	0.23	12.2	1.5	16.3	67.5	–	15.0	17.2	67.8	Fernandes et al. (2014b)
Russula olivacea	0.4	0.3	16.1	2.8	26.0	50.2	0.1	21.8	27.4	50.8	Grangeia et al. (2011)
Russula virescens	–	–	17.3	7.2	40.3	29.2	–	28.8	41.5	29.7	Leal et al. (2013)
Sprassis crispa	–	2.0	10.4	1.7	49.0	31.3	ND	19.0	49.7	31.3	Kavishree et al. (2008)
Suillus granulatus	ND	0.2	12.0	3.3	34.2	46.6	0.2	17.2	35.7	47.1	Pedneault et al. (2006)
Suillus granulatus	Traces	0.1	0.4	0.4	62.8	35.4	ND	1.3	63.3	35.4	Ribeiro et al. (2009)
from Portugal	Traces	0.2	9.6	3.2	24.6	57.1	0.4	15.4	26.6	58.0	Reis et al. (2014b)
from Serbia	Traces	0.2	9.6	2.7	20.1	64.0	0.2	14.3	21.3	64.4	Reis et al. (2014b)
Suillus grevillei	Traces	0.2	8.9	1.5	43.9	40.1	0.6	12.6	45.5	41.9	Pedneault et al. (2006)
Suillus luteus	ND	0.1	11.9	0.6	ND	67.1	1.4	–	–	–	Karliński et al. (2007)
Suillus luteus	–	–	7.6	4.1	37.8	44.1	0.7	–	–	–	Dembitsky et al. (2010)
Suillus variegatus	–	–	12.7	3.5	42.0	37.4	–	18.1	44.2	37.7	Pereira et al. (2012)
Terfezia boudieri	–	–	23.7	4.0	10.4	44.3	2.1	–	–	–	Dundar et al. (2012)
Termitomyces robustus	0.1	0.9	20.9	4.0	9.5	59.2	0.2	30.1	10.3	59.6	Obodai et al. (2014)
Tricholoma imbricatum	0.1	0.2	7.4	4.1	51.5	33.0	0.2	14.9	51.8	33.3	Heleno et al. (2009)

(Continued)

Table 2.5 Proportion of major fatty acids (% of total fatty acids) in selected mushroom species (Continued)

Species	Lauric acid	Myristic acid	Palmitic acid	Stearic acid	Oleic acid	Linoleic acid	Linolenic acid	SFA	MUFA	PUFA	Reference
Tricholoma portentosum	0.4	1.2	7.6	3.4	58.0	27.9	–	–	–	–	Díez and Alvarez (2001)
	–	0.1	5.6	2.3	58.4	30.9	0.4	9.6	59.0	31.4	Barros et al. (2007a)
Tricholoma terreum	0.2	0.3	10.1	1.8	56.7	29.7	–	–	–	–	Díez and Alvarez (2001)
Xerocomus badius	ND	0.15	20.2	ND	2.2	70.9	Traces	–	–	–	Karliński et al. (2007)
	–	–	15.1	2.1	36.5	38.2	1.6	–	–	–	Dembitsky et al. (2010)
Xerocomus subtomentosus	0.2	0.4	16.3	1.7	31.7	42.2	0.3	22.4	34.4	43.2	Pedneault et al. (2006)
	ND	0.2	18.5	0.1	7.3	66.7	ND	–	–	–	Karliński et al. (2007)
Cultivated species											
Agaricus bisporus white	–	–	11.9	3.1	1.1	77.7	0.1	20.3	1.4	78.3	Reis et al. (2012)
brown	–	–	11.1	3.0	1.2	79.4	0.1	18.4	1.8	79.8	Reis et al. (2012)
	0.1	0.6	14.0	6.5	20.6	43.9	3.3	30.3	21.7	48.0	Stojković et al. (2014)
	0.1	0.6	15.4	3.7	14.9	60.4	0.9	23.1	15.6	61.6	Glamočlija et al. (2015)
Agaricus subrufescens	Traces	0.2	11.3	3.1	1.0	74.5	0.2	22.1	1.5	76.4	Stojković et al. (2014)

Species											Reference
Auricularia auricula-judae	0.1	0.6	18.8	11.3	27.2	34.6	1.6	35.8	27.7	36.5	Obodai et al. (2014)
Coprinus comatus	–	–	8.6	2.1	36.4	50.5	–	11.8	36.8	51.4	Stojković et al. (2013)
Flammulina velutipes	–	–	11.0	2.0	5.7	45.4	28.8	18.5	7.2	74.3	Reis et al. (2012)
Lentinula edodes	–	–	10.3	1.6	2.3	81.1	0.1	15.1	2.9	82.0	Reis et al. (2012)
Lentinula squarrosulus	0.1	0.6	18.8	3.3	8.3	62.9	1.1	26.6	9.1	64.3	Obodai et al. (2014)
Pleurotus eryngii	–	–	12.8	1.7	12.3	68.9	0.3	17.4	13.1	69.4	Reis et al. (2012)
Pleurotus ostreatus	–	–	11.2	3.0	12.3	68.8	0.3	17.0	13.6	69.4	Reis et al. (2012)
	0.3	1.0	11.2	2.5	9.5	68.1	0.2	20.2	10.8	69.1	Fernandes et al. (2015a)
Pleurotus sajor-caju	<0.1	0.4	14.3	4.4	18.3	58.5	0.2	21.9	18.9	59.2	Obodai et al. (2014)
	0.1	0.6	18.8	4.6	22.6	48.6	0.2	26.6	24.2	49.2	Obodai et al. (2014)
Pleurotus tuber-regium	0.3	1.1	21.2	9.6	21.2	32.3	1.1	42.2	24.1	33.7	Obodai et al. (2014)

For fatty acid characteristics see Appendix III. SFA, saturated fatty acids; MUFA, monounsaturated fatty acids; PUFA, polyunsaturated fatty acids. ND, the content below the level of detection limit.

dried ones (see Section 3.1.2). Sporadic and fragmentary information has been published on the occurrence of conjugated linoleic acid (CLA; see Appendix III) in *A. bisporus*. CLA is the term for a group of 18-carbon acids with two double bonds in positions 9,11 or 10,12 and one *trans*-configuration, or for prevailing 9-*cis*, 11-*trans*-linoleic acid. The latter compound occurs mainly in fats of ruminants and is called rumenic acid. It has been known as a chemopreventive agent against several types of cancer.

In sliced fruit bodies of *P. ostreatus* irradiated with UV-B light (310–320 nm) there has decreased content of only linoleic acid, but no significant changes of the main fatty acids were observed in intact fruit bodies (Krings and Berger, 2014). In a series of experiments with several mushroom species, gamma irradiation affected the proportion of saturated, monounsaturated, and polyunsaturated fatty acids to a lesser extent than did oven-drying or freezing (Fernandes et al., 2013a,b,c, 2014a).

Overall, with their very low lipid content, only approximately 2–3 g $100\,g^{-1}$ DM or approximately 0.2–0.3 g $100\,g^{-1}$ FM, and a low proportion of both saturated and ω–3 fatty acids, mushrooms rank among food items with a marginal nutritional role of lipids.

2.4 CARBOHYDRATES AND DIETARY FIBER

As indicated in Tables 2.1 and 2.2, carbohydrates form the main proportion of mushroom DM. Groups of carbohydrates comprise numerous compounds. Monosaccharides and their derivatives, in mushrooms particularly polyol (or alcoholic sugar) mannitol, oligosaccharides formed by 2–10 monosaccharides linked with glycosidic bonds, and various polysaccharides (glycans) contain from tens to thousands of linked monosaccharides.

Generally, the composition of mushroom carbohydrates, from monosaccharides and oligosaccharides, commonly named sugars, to polysaccharides, differs from that of plants. Mushrooms contain glycogen as an energy reserve instead of starch and distinct structural polysaccharides. Different opinions exist regarding cellulose occurrence in mushrooms. While some authors admit its presence, Bauer Petrovska (2002) did not find cellulose in fiber isolates of 53 mushroom species. Nevertheless, cellulose is a component of fungal cell walls (eg, of the yeast *Saccharomyces cerevisiae*).

2.4.1 Sugars

Available data on the main monosaccharides (arabinose, fructose, and glucose), oligosaccharides (melezitose and trehalose), and polyol mannitol occurring in mushrooms are collated in Table 2.6. Total sugar contents vary widely, between 5 and 25 g $100\,g^{-1}$ DM (ie, approximately 0.5–2.5 g $100\,g^{-1}$ FM), with several outlying data. Mannitol and trehalose (more accurately α, α-trehalose formed by two molecules of α-D-glucopyranose bound by $1\rightarrow1$ glycosidic bond) occur at the highest level by far. However, their contents vary widely among species. As may be seen from limited data for several species (eg, for *Coprinus comatus* or *B. edulis*) originating from various laboratories, great variability also occurs within a species. Information on other sugars, such as xylose, mannose, sucrose and maltose, deoxysugars fucose, and rhamnose or polyol arabitol, has been very sporadic and mostly negligible contents have been reported. The sugars are water-soluble and partly contribute to mushroom taste (see Section 3.1.1). Mannitol is synthesized through the reduction of fructose by mannitol dehydrogenase. It participates in volume growth and firmness of fruit bodies. Such roles are supported by an observed increase of mannitol level in mature fruit bodies of several cultivated species, as compared with the immature ones. Mannitol is of limited sweetness, with approximately half the calories of common sugars, and is poorly absorbed by the human body. It does not raise insulin levels as sucrose does.

Maldigestion of trehalose in some individuals can cause abdominal symptoms similar to those of lactose maldigestion and intolerance. One of the factors is low activity of trehalase in the small bowel. The activity is significantly reduced in untreated celiac disease. Nevertheless, a very low frequency of such disease was observed within the UK population (Murray et al., 2000).

Table 2.6 Contents of soluble sugars and polyols (g 100 g^{-1} dry matter) in selected mushroom species

Species	Arabinose	Fructose	Glucose	Mannitol	Melezitose	Trehalose	Total sugars	Reference
Wild-growing species								
Agaricus albertii	—	0.48	—	4.78	—	0.70	5.0	Reis et al. (2014a)
Agaricus arvensis	—	—	—	6.5	—	0.4	6.9	Barros et al. (2007a)
Agaricus bitorquis	—	0.40	—	7.04	—	2.27	11.2	Glamočlija et al. (2015)
Agaricus campestris	—	ND	—	16.9	—	3.62	20.5	Pereira et al. (2012)
Agaricus campestris	—	0.29	—	5.59	—	0.63	6.51	Glamočlija et al. (2015)
Agaricus macrosporus	—	2.65	—	4.98	—	1.15	8.78	Glamočlija et al. (2015)
Agaricus silvaticus	—	—	—	2.65	0.38	0.25	8.72	Barros et al. (2008b)
Agaricus silvicola	—	—	—	6.09	0.47	0.66	7.79	Barros et al. (2008b)
Agaricus urinascens var. excellens	—	0.60	—	0.77	—	0.14	1.5	Reis et al. (2014a)
Amanita cesarea	—	ND	—	0.30	—	0.58	0.88	Fernandes et al. (2015c)
Amanita curtipes	—	2.3	—	3.9	—	8.9	15.1	Fernandes et al. (2015c)
Armillariella mellea	0.78	—	—	5.45	—	9.33	15.7	Vaz et al. (2011)
Auricularia auricula-judae	—	ND	—	0.68	—	2.52	3.87	Obodai et al. (2014)

Boletus aestivalis	ND	ND	—	2.93	—	3.92	6.85	Heleno et al. (2011)
Boletus appendiculatus	ND	—	—	1.34	—	4.65	6.0	Heleno et al. (2011)
Boletus armeniacus	—	10.5	—	23.6	—	5.62	39.7	Pereira et al. (2012)
Boletus edulis	—	—	—	3.46	0.29	9.71	13.5	Barros et al. (2008b)
	ND	—	—	2.45	—	12.4	14.85	Heleno et al. (2011)
	—	0.41	1.57	1.13	—	17.7	20.9	Fernandes et al. (2013b)
	—	0.08	ND	0.15	—	3.2	4.0	Fernandes et al. (2014b)
Boletus erythropus	—	1.72	—	27.9	—	4.84	34.5	Grangeia et al. (2011)
Boletus regius	—	14.0	—	6.25	—	0.66	20.9	Leal et al. (2013)
Calocybe gambosa	—	—	—	0.27	0.85	8.01	9.13	Barros et al. (2008b)
	ND	—	—	0.29	—	7.96	8.25	Vaz et al. (2011)
Calvatia utriformis	—	ND	—	ND	—	0.40	0.40	Grangeia et al. (2011)
Cantharellus cibarius	—	—	—	13.9	ND	11.2	25.1	Barros et al. (2008a)
	—	—	—	8.33	ND	6.12	14.5	Barros et al. (2008b)
Clitocybe odora	ND	—	—	0.59	—	7.77	8.36	Vaz et al. (2011)
Coprinus comatus	ND	—	—	0.40	—	42.8	43.2	Vaz et al. (2011)
	—	ND	—	1.84	—	5.41	7.25	Stojković et al. (2013)

(*Continued*)

Table 2.6 Contents of soluble sugars and polyols (g 100 g^{-1} dry matter) in selected mushroom species (Continued)

Species	Arabinose	Fructose	Glucose	Mannitol	Melezitose	Trehalose	Total sugars	Reference
Craterellus cornucopioides	–	–	–	10.7	ND	0.11	10.8	Barros et al. (2008b)
Flammulina velutipes	–	ND	–	5.98	–	15.1	21.1	Pereira et al. (2012)
Hericium coralloides	6.25	–	–	3.86	–	0.68	–	Heleno et al. (2015)
Hericium erinaceus	17.5	–	–	5.63	–	0.54	–	Heleno et al. (2015)
Hydnum repandum	–	ND	ND	13.0	–	4.4	17.4	Fernandes et al. (2013c)
Laccaria laccata	ND	–	–	0.64	–	5.81	6.45	Heleno et al. (2009)
Lactarius deliciosus	–	–	–	13.7	–	2.7	16.4	Barros et al. (2007a)
	–	0.18	–	12.0	–	1.4	–	Fernandes et al. (2013b)
Laetiporus sulphureus	–	0.46	–	3.54	–	4.0	8.0	Petrović et al. (2014)
Lepista nuda	–	–	–	0.8	ND	12.0	12.8	Barros et al. (2008a)
Leucopaxillus giganteus	–	–	–	1.85	–	6.6	8.45	Barros et al. (2007a)
Lycoperdon perlatum	–	–	–	0.2	ND	2.6	2.8	Barros et al. (2008a)
Macrolepiota procera	–	0.06	–	5.2	1.24	9.1	15.7	Fernandes et al. (2013a)

Marasmius oreades	—	—	—	2.42	0.55	10.5	13.5	Barros et al. (2008b)
Morchella esculenta	—	0.71	—	11.5	—	3.41	15.6	Heleno et al. (2013)
	—	ND	—	1.08	—	5.34	6.42	Heleno et al. (2013)
Pleurotus eryngii	—	ND	—	1.40	—	14.2	15.6	Reis et al. (2014a)
Ramaria botrytis	—	—	—	11.7	0.2	2.0	13.9	Barros et al. (2008a)
Russula cyanoxantha	—	0.34	—	16.2	—	1.64	18.2	Grangeia et al. (2011)
Russula delica	—	ND	2.37	4.28	—	2.83	10.3	Fernandes et al. (2014b)
Russula olivacea	—	0.23	—	15.3	—	0.71	16.2	Grangeia et al. (2011)
Russula virescens	—	ND	—	10.9	—	0.2	11.1	Leal et al. (2013)
Suillus granulatus from Portugal	—	4.5	—	3.33	—	4.86	12.7	Reis et al. (2014b)
from Serbia	—	7.02	—	3.18	—	2.57	12.8	Reis et al. (2014b)
Suillus variegatus	—	ND	—	ND	—	4.85	4.85	Pereira et al. (2012)
Termitomyces robustus	—	ND	—	4.71	—	9.92	14.63	Obodai et al. (2014)
Tricholoma imbricatum	ND	—	—	10.5	—	6.56	17.1	Heleno et al. (2009)
Tricholoma portentosum	—	—	—	1.0	—	21.0	22.0	Barros et al. (2007a)

(Continued)

Table 2.6 Contents of soluble sugars and polyols (g 100 g^{-1} dry matter) in selected mushroom species (Continued)

Species	Arabinose	Fructose	Glucose	Mannitol	Melezitose	Trehalose	Total sugars	Reference
Cultivated species								
Agaricus bisporus								
white	—	0.03	—	5.6	—	0.16	5.8	Reis et al. (2012)
brown	—	0.04	—	4.0	—	0.22	4.26	Reis et al. (2012)
	—	0.97	—	12.1	—	1.21	14.6	Pei et al. (2014)
	—	—	—	14.9	—	ND	14.9	Stojković et al. (2014)
	—	ND	—	11.3	—	0.60	11.9	Glamočlija et al. (2015)
Agaricus subrufescens dried powder	—	—	2.76	7.94	—	2.98	18.5	Tsai et al. (2008)
	—	0.27	—	60.9	—	5.7	66.9	Carneiro et al. (2013)
	—	0.13	Traces	15.2	—	1.22	16.5	Li et al. (2014)
	—	—	—	13.8	—	0.49	14.3	Stojković et al. (2014)
Coprinus comatus	—	0.11	—	1.41	—	8.75	10.4	Stojković et al. (2013)
Flammulina velutipes	—	Traces	Traces	8.39	—	17.4	25.8	Li W. et al. (2014)
	—	4.60	—	0.97	—	2.63	8.2	Reis et al. (2012)
Hypsizigus marmoreus	0.66	0.27	—	2.26	—	0.98	4.17	Lee et al. (2009)

							Reference
white	–	0.36	–	–	–	14.3	21.5	Wu et al. (2015)
Lentinula edodes	–	–	2.86	8.38	–	2.92	14.2	Yang et al. (2001)
	–	0.69	–	10.0	–	3.38	14.1	Reis et al. (2012)
caps	Traces	0.67	4.34	56.7	–	36.6	111.6	Chen et al. (2015)
stipes	0.39	0.43	1.57	34.4	–	40.1	125.0	Chen et al. (2015)
dried powder	1.79	ND	–	23.3	–	13.2	38.3	Carneiro et al. (2013)
Lentinula squarrosulus	–	0.42	–	1.15	–	12.7	14.3	Obodai et al. (2014)
	–	–	6.64	2.58	–	0.29	–	Zhou et al. (2015)
Oudemansiella submucida	–	–	0.97	2.90	–	ND	–	Zhou et al. (2015)
Pleurotus eryngii	–	ND	ND	1.5	–	30.2	31.7	Li W. et al. (2014)
	–	0.03	–	0.60	–	8.01	8.64	Reis et al. (2012)
	–	0.05	0.31	1.17	–	30.1	31.5	Li et al. (2015)
Pleurotus ostreatus	–	–	1.06	0.36	–	0.27	1.69	Yang et al. (2001)
	–	ND	1.43	–	–	0.30	–	Kim et al. (2009)
	–	0.01	–	0.54	–	4.42	4.97	Reis et al. (2012)
	–	0.30	–	0.87	–	12.7	13.9	Obodai et al. (2014)
Pleurotus sajor-caju	–	0.32	–	1.99	–	6.61	8.92	Obodai et al. (2014)
Pleurotus tuber-regium	–	ND	–	0.35	–	1.50	1.85	Obodai et al. (2014)
Tremella aurantialba	–	–	0.63	0.89	–	1.66	–	Zhou et al. (2015)

ND, content below detection limit. In several species, total content is higher than sum of the values for individual sugars due to the occurrence of other sugars. Total contents are rounded off to three significant digits.

Table 2.7 Mannitol and trehalose contents (g 100 g^{-1} dry matter) in preserved and cooked mushrooms

Species	Mannitol			Trehalose		
	Dried	Frozen	Cooked	Dried	Frozen	Cooked
Lactarius deliciosus	15.4	13.9	10.2	0.9	3.5	2.2
Macrolepiota mastoidea	7.8	4.6	7.5	4.1	5.1	ND
Macrolepiota procera	4.7	6.5	2.3	2.9	7.6	1.2
Sarcodon imbricatus	19.6	25.3	11.8	6.0	5.0	3.4

Adapted from Barros et al. (2007b). Copyright 2007, American Chemical Society. With permission.
Dried at 40°C; frozen at −20°C; cooked with olive oil, onions, and salt. ND, content below detection limit.

Data on changes in sugar content and composition caused by different mushroom storage and processing conditions have been limited. Barros et al. (2007b) compared effects of drying at 40°C, freezing to −20°C, and cooking with olive oil, onions, and table salt on the changes in mannitol and trehalose contents in four species (Table 2.7). Unfortunately, initial data prior to preservation and cooking were not given. Cooking caused the highest losses of both the compounds, and the effects of drying and freezing were not significantly different.

Trehalose was the main sugar in *P. eryngii*, accounting for more than 94% of total soluble sugars. Five drying methods, namely ambient air drying, hot air drying, freeze, or vacuum or microwave drying, caused only limited changes in trehalose content. Similar relative stability was observed in mannitol content (Li et al., 2015).

Different data for *Macrolepiota procera* were reported by Fernandes et al. (2013a). The mannitol contents were 5.2, 11.4, 4.9, and 5.1 g 100 g^{-1} DM in fresh, dried (at 30°C), frozen, and gamma-irradiated fruit bodies, respectively. The respective values for trehalose were 9.1, 6.8, 3.0, and 10.2 g 100 g^{-1} DM. This and further reports of a similar nature (Fernandes et al., 2013b,c, 2014a) demonstrated that gamma irradiation maintained the profile of sugars (similarly as of other components) to a greater extent than traditional preservation by drying or freezing.

2.4.2 Polysaccharides and Dietary Fiber

The major types of mushroom polysaccharides include chitin, glucans, and heteroglycans. The usual content of reserve polysaccharide glycogen, an alpha-glucan also called "animal starch," is 5–15% of DM (ie, approximately 0.5–1.5% of fresh weight) (Dikeman et al., 2005). Some evidence exists that glycogen together with disaccharide trehalose supply a considerable proportion of carbon for growth of fruit bodies (eg, for mannitol synthesis). Their peak levels in mycelium are related to the productivity of the emerging flush. Glycogen is widely consumed in human nutrition, mainly in meat and liver. Its low intake from mushrooms thus seems to be nutritionally unimportant.

Great interest has been focused on specific mushroom structural polysaccharides, beta-glucans, due to their positive health effects. Information on these compounds are given in Section 4.2.

The main components of the mushroom cell wall are structural polysaccharides, which account for up to 80–90% of cell wall DM. Among them, chitin is characteristic for mushrooms. It is a water-insoluble polymer of β-(1→4)-branched N-acetyl-D-glucosamine units. Partial deacetylation of chitin yields chitosan. The reported content in *T. aestivum* was $10.6\pm 3.6\,g\ 100\,g^{-1}$ DM (Kruzselyi and Vetter, 2014), and in *A. bisporus* and *Lentinula edodes* it was $5–10\,g\ 100\,g^{-1}$ DM (Manzi et al., 2001; Vetter, 2007), whereas only $2–4\,g\ 100\,g^{-1}$ DM has been reported for *P. ostreatus* (Vetter, 2007). These contents are somewhat higher than the mean levels determined by Nitschke et al. (2011) in six cultivated species: 9.60, 4.69, 3.16, 1.87, 0.76, and $0.39\,g$ $100\,g^{-1}$ DM (expressed as chitosan) in *F. velutipes*, *A. bisporus*, *P. eryngii*, *L. edodes*, *P. ostreatus*, and *Hypsizygus tessulatus*, respectively. Nevertheless, it is not reliable to compare chitin contents determined by various analytical methods. Chitin is insoluble in most solvents and its direct determination is thus difficult. It can be quantified indirectly as chitosan or N-acetylglucosamine.

According to Vetter (2007), significantly higher chitin content was observed in caps than in stipes of *A. bisporus*. Moreover, chitin content in the three aforementioned cultivated species seems to be characteristic for a species and independent for varieties. Its content is lower in wood-rotting species than in saprobic and ectomycorrhizal ones. Chitin is indigestible for humans, thus forming an important part of mushroom insoluble dietary fiber. It seems to be responsible for decreasing the cholesterol pool in experimental animals ingesting a *P. ostreatus* diet. Chitin apparently decreases the digestibility of other mushroom components.

Dietary fiber is the sum of intrinsic carbohydrates indigestible in the small intestine. A further component in plant foods is lignin of polyphenolic nature, which is absent in mushrooms. Total dietary fiber (TDF) consists of soluble and insoluble parts. Soluble fiber forms a gel with water, which increases viscosity of foods and chyme. Digestion is slowed, which can lower postprandial blood glucose, insulin, and cholesterol levels. The insoluble fraction softens stools and increases stool bulk, contributing to good health conditions of the colon.

It should be noticed here that the analytical methods used for fiber determination give different results and the reported data should be interpreted cautiously. The reported proportion of TDF in 12 wild-growing mushroom species varied widely between 5.1% and 40.0% in DM, with mean and median values of 26% and 30.5%, respectively (Wang et al., 2014). More narrow TDF variability, with a usual proportion of approximately 30% in DM, was observed in wild mushrooms collected in India (Table 2.8). Insoluble dietary fiber amounted, on average, to 60% of TDF (Nile and Park, 2014). Such data fit well with available information for cultivated species. The TDF proportions of 25–30% have been the most frequent. Lower values of approximately 20% have been reported in *T. aestivum* (Kruzselyi and Vetter, 2014), in cultivated *Agaricus* spp. (Mattila et al., 2002; Tsai et al., 2008; Lee et al., 2009; Khan and Tania, 2012), and in cultivated *P. eryngii* (Cui et al., 2014). An even lower level of 5.6% in DM was determined by Kovács and Vetter (2015) in wood-decaying

Table 2.8 Content of total, insoluble, and soluble dietary fiber (g 100 g^{-1} dry matter) determined by AOAC enzymatic-gravimetric method in selected wild mushroom species

Species	Total dietary fiber	Insoluble dietary fiber	Soluble dietary fiber
Agaricus bisporus	31 ± 1.4	14 ± 1.9	2
Auricularia polytricha	32 ± 2.3	17 ± 1.8	3
Boletus edulis	28 ± 2.1	13 ± 2.3	3
Cantharellus cibarius	36 ± 2.1	20 ± 2.4	3
Cantharellus clavatus	36 ± 1.7	12 ± 1.6	4
Helvella crispa	28 ± 2.4	21 ± 3.2	3
Hericium erinaceus	30 ± 2.3	14 ± 1.2	3
Hydnum repandum	33 ± 1.4	17 ± 1.8	3
Lactarius sanguifluus	24 ± 1.1	18 ± 1.5	3
Morchella conica	30 ± 1.8	16 ± 3.2	2
Pleurotus sajor-caju	27 ± 2.1	12 ± 2.3	2
Sparassis crispa	32 ± 1.6	15 ± 1.5	4

Adapted from Nile and Park (2014). With permission.

L. sulphureus. On the contrary, Heleno et al. (2015) determined TDF to be more than 40% in DM and determined the proportion of indigestible fiber to be more than 90% in *Hericium coralloides* and *H. erinaceus.*

Cooking increased the TDF proportion in DM as compared with raw mushrooms. The proportion of chitin ranged between 8% and 32% of TDF in several tested species, with *Boletus* spp. being richest (Manzi et al., 2001, 2004). Preservation treatment of air-dried *B. edulis* and *M. procera* with electron beam irradiation up to a dose of 10 kGy had only limited effect on both insoluble and soluble fiber of *B. edulis*, whereas increasing doses decreased the content of the insoluble fraction and increased the soluble fraction in *M. procera*. However, both irradiated species remain a good source of dietary fiber (Fernandes et al., 2015b).

Heteroglycans are further components of dietary fiber. They have been characterized insufficiently. An illustration of their components is seen from the data in Table 2.9. Glucose and glucosamine prevail considerably, followed by mannose and uronic acids.

Table 2.9 Proportion of total dietary fiber (TDF; % of dry matter) and monosaccharide composition of total dietary fiber (% of total polysaccharides). The values were determined by Uppsala Methods

Species	TDF	Rhamnose	Xylose	Mannose	Galactose	Glucose	Glucosamine	Uronic acids
Lentinula edodes								
Caps	32.4	ND	1.92	5.21	3.30	70.8	15.0	3.75
Stipes	40.0	0.65	1.47	5.23	1.77	73.9	13.9	3.12
Pleurotus sajor-caju								
Caps	29.1	ND	1.14	5.12	3.47	78.2	7.90	4.17
Stipes	30.9	ND	0.62	3.77	2.01	82.8	6.50	4.34
Volvariella volvacea								
Caps	22.8	ND	ND	6.22	4.48	70.5	13.9	4.91
Stipes	24.3	ND	ND	6.01	4.11	72.0	13 1	4.73

Adapted from Cheung (1996). Copyright 1996, American Chemical Society. With permission.
ND, content below detection limit.

The recommended daily intake of TDF is 30 g for an adult. Mushrooms are thus a suitable source of dietary fiber, particularly its prevailing insoluble part, which is commonly insufficient in human nutrition. One portion comprising 100 g of fresh mushrooms provides approximately 10% of the recommended daily intake.

2.5 MAJOR MINERALS

Crude ash consists of seven major elements, which quantitatively highly prevail, and of tens of trace elements commonly occurring at the level up to 50 mg/kg of FM for each of them. Information on trace elements is given in Section 3.9 and information on radioactive ones is given in Section 5.7.

Mushrooms usually contain 5–12 g of ash 100 g^{-1} DM (ie, approximately 0.5–1.2 g in 100 g FM). Such an extent is typical for cultivated species (Table 2.2), whereas contents more than 20 g 100 g^{-1} DM were reported for several wild-growing species (Table 2.1). This probably results from the greater variability of substrates in nature. Nevertheless, as shown by data of Table 2.10, variability in ash contents (except for *Craterellus cornucopioides*) seems to be lower than in other major components (eg, in crude protein as shown in Table 2.3).

The usual major mineral elements contents are given in Table 2.11 and information on variability of three elements are given in Table 2.12. The levels of cultivated species are usually in the lower ranges. Potassium is by far the main major mineral element, followed by phosphorus. The sodium and calcium contents are low. The vast data for more than 400 wild species, not exceeded until now, were published by the laboratory of Professor Ruth Seeger from the University of Würzburg, Germany, in the 1970s and 1980s. Data on four alkaline metals have been supplemented until now, and information on sulfur and particularly chlorine has been limited.

Potassium is not distributed evenly within fruit bodies. Its content decreases in the following order: cap > stipe > spore-forming part > spores. Potassium levels are between 20- and

Table 2.10 Variability of ash content in several mushroom species

Species	Number of samples	Mean (% of dry matter)	Standard deviation (% of dry matter)	Relative standard deviation (%)
Armillariella mellea	6	15.15	1.44	9.5
Boletus edulis	10	7.45	0.68	9.1
Cantharellus cibarius	8	19.96	4.60	23.0
Clitocybe nebularis	6	10.89	1.09	10.0
Craterellus cornucopioides	6	16.90	8.10	47.9
Leccinum scabrum	9	11.53	1.22	10.6
Marasmius oreades	3	11.71	0.51	4.35
Xerocomus subtomentosus	7	10.54	0.15	1.4

Adapted from Vetter (1993a). With permission.

Table 2.11 Usual content (mg $100\,g^{-1}$ dry matter) of major mineral elements in wild-growing mushrooms

Element	Usual content
Sodium	10–40
Potassium	2000–4000
Calcium	20–100
Magnesium	80–180
Phosphorus	500–1000
Sulfur	100–300
Chlorine	100–600

40-fold higher in fruiting bodies than in underlying substrates. Bioaccumulation has not been observed for sodium and calcium. Magnesium contents in fruit bodies were even lower than in substrates.

The total phosphorus content of 825 samples was reported by Quinche (1997). Significantly differing mean levels were 610 and 1370 mg $100\,g^{-1}$ DM in 51 ectomycorrhizal and 42 saprobic

Table 2.12 Variability in three major minerals (mg 100 g^{-1} dry matter)

Species	n	Phosphorus			Potassium			Calcium		
		Mean	SD	RSD	Mean	SD	RSD	Mean	SD	RSD
Armillariella mellea	6	1100	140	13.2	3890	420	10.8	34	4	11.7
Boletus edulis	10	720	140	19.4	2870	460	16.0	21	2	9.5
Cantharellus cibarius	8	670	100	14.9	3900	590	15.1	27	6	22.2
Clitocybe nebularis	6	1670	140	8.3	3360	200	5.9	26	4	15.3
Craterellus cornucopioides	6	630	50	7.9	3710	530	14.2	24	4	16.6
Leccinum scabrum	9	690	120	17.4	3150	180	5.7	18	6	33.3
Marasmius oreades	3	690	207	30.0	2080	690	33.3	25	6	24.0
Xerocomus subtomentosus	7	840	125	14.8	3110	340	10.9	23	1	4.3

Adapted from Vetter (1993a). With permission.
n, number of samples; SD, standard deviation (mg 100 g^{-1} dry matter); RSD, relative standard deviation (%).

species, respectively. Usual contents were 500–700 mg $100\,g^{-1}$ DM in the Boletaceae family, whereas usual levels were 1300–2300 mg $100\,g^{-1}$ DM in the genus *Lepista*. The tested species accumulated phosphorus in their bodies at contents between 10- and 50-fold higher than in the underlying substrates.

Mean total sulfur content 220 ± 110 mg $100\,g^{-1}$ DM was found in eight widely consumed wild species, with ranges between 90 and 440 mg $100\,g^{-1}$ DM for *A. rubescens* and *Xerocomus chrysenteron*, respectively. Although the differences in sulfur content were highly significant among the tested species, insignificant differences were observed between caps and stipes (Rudawska and Leski, 2005). In a more recent article on 27 wild species from Canada, there was a reported mean total sulfur content of 263 ± 91 mg $100\,g^{-1}$ DM. The minimum and maximum levels of 73 and 1387 mg $100\,g^{-1}$ DM were observed in *Hydnum repandum* and *B. edulis*, respectively (Nasr et al., 2012).

Recent partial reviews on major mineral elements are available for Chinese wild species (Wang et al., 2014) and species of the genus *Pleurotus* (Khan and Tania, 2012). The collated data were inserted into the values of Table 2.10. The latter review reports surprisingly high sodium contents in *Pleurotus flabellatus* and a low level of potassium in *Pleurotus tuber-regium*.

Çağlarirmak (2009) studied changes in ash and five major elements during three flushes of cultivated brown *A. bisporus* (portobello). Although ash, phosphorus, and sodium contents did not change significantly, calcium and magnesium contents decreased, and that of potassium increased between the first and third flush. Similarly, as mentioned for potassium, the major elements are distributed unevenly within a fruit body. The contents are usually higher in caps than in stipes; however, an inverse relation was observed for calcium and sodium in *Agaricus subrufescens* (Györfi et al., 2010).

Generally, ash content of mushrooms is higher than or comparable with that of most vegetables. This particularly applies to phosphorus and potassium, whereas sodium and calcium levels are often lower than that in vegetables. Nevertheless, information on the bioavailability of the essential major elements from mushrooms has been lacking.

2.6 VITAMINS AND PROVITAMINS

The contents of vitamins and provitamins as their precursors are discussed according to the traditional classifications of fat-soluble and water-soluble groups.

2.6.1 Fat-Soluble Vitamins and Provitamins

Data are presented in Table 2.13. Beta-carotene contents, as the most potent precursor of vitamin A, are low, commonly between non-detectable level and 0.6 mg $100 g^{-1}$ DM. Generally, the occurrence of carotenoids in mushrooms is considerably lower than in plants (see Section 3.2). Information on mushroom beta-carotene bioavailability in the human intestinal tract has been lacking. Thus, the nutritional contribution of mushroom beta-carotene seems to be marginal.

On the contrary, mushroom ergosterol, the provitamin of vitamin D_2 (ergocalciferol), is nutritionally significant. Its content is highest among mushroom sterols (mycosterols) (see Section 3.5) and varies widely from tens to several hundreds of mg $100 g^{-1}$ DM (Table 2.13). Carvalho et al. (2014) reported a somewhat higher mean content of 103 mg $100 g^{-1}$ FM (ie, approximately 1000 mg $100 g^{-1}$ DM) in 12 edible wild species. The content of free ergosterol in both white and brown cultivated *A. bisporus* varied, with a range of 204–482 mg $100 g^{-1}$ DM. A higher level was observed in early growth stages, but the level decreased as the fruit bodies grew. Ergosterol was distributed evenly between caps and stems during early developmental stages but increased in caps after maturation (Shao et al., 2010).

Ergosterol occurs in two main forms, free and esterified with acids as ergosteryl esters. Free molecules participate in fluidity, permeability, and integrity of the cell membranes while the esters

Table 2.13 Content of fat-soluble provitamin ergosterol (mg 100 g^{-1} dry matter) and tocopherols (µg 100 g^{-1} dry matter) in selected mushroom species

Species	Ergosterol	Tocopherols					Reference
		Alpha-	Beta-	Gamma-	Delta-	Total	
Wild-growing species							
Agaricus albertii	–	ND	ND	ND	ND	ND	Reis et al. (2014a)
Agaricus arvensis	–	7.0	115	ND	ND	122	Barros et al. (2008c)
Agaricus bitorquis	–	5.1	ND	11	18.8	34.9	Glamočlija et al. (2015)
Agaricus campestris	–	10	30	40	30	110	Pereira et al. (2012)
Agaricus macrosporus	–	6.4	ND	110	ND	116.4	Glamočlija et al. (2015)
Agaricus romagnesii	–	4.1	ND	26.9	ND	31.0	Glamočlija et al. (2015)
	–	ND	129	ND	ND	129	Barros et al. (2008c)
Agaricus silvicola	–	130	193	ND	ND	323	Barros et al. (2008c)
Agaricus urinascens	–	48.7	67.7	ND	ND	116.4	Barros et al. (2008b)
var. *excellens*	–	ND	ND	ND	ND	ND	Reis et al. (2014a)
Amanita caesarea	231	–	–	–	–	–	Barreira et al. (2014)
Amanita curtipes	–	6.4	ND	100	34	140	Fernandes et al. (2015c)
Armillariella mellea	–	4.0	ND	40.0	4.1	48	Fernandes et al. (2015c)
Boletus aestivalis	–	9.1	ND	58.0	ND	67.1	Vaz et al. (2011)
Boletus appendiculatus	–	10	ND	2520	ND	2530	Heleno et al. (2011)
	–	14	ND	566	29	609	Heleno et al. (2011)

							Reference
Boletus edulis	241	–	–	–	–	–	Teichmann et al. (2007)
	–	32	890	142	ND	1064	Barros et al. (2008b)
	–	12	ND	517	51	580	Heleno et al. (2011)
	234	–	–	–	–	–	Barreira et al. (2014)
	400	–	–	–	–	–	Villares et al. (2014)
Boletus erythropus	–	17	–	20	57	94	Fernandes et al. (2014b)
Boletus regius	–	1.5	ND	17.6	ND	19.1	Grangeia et al. (2011)
	–	361	ND	403	ND	764	Leal et al. (2013)
Calocybe gambosa	–	6.0	20	14	ND	40.0	Barros et al. (2008b)
	–	15.8	ND	119	24.7	160	Vaz et al. (2011)
	–	–	–	–	–	–	Villares et al. (2014)
Calvatia utriformis	361	8.7	ND	56.4	ND	65.1	Grangeia et al. (2011)
Cantharellus cibarius	348	–	–	–	–	–	Teichmann et al. (2007)
	–	12	3.0	3.0	ND	18.0	Barros et al. (2008b)
	129	–	–	–	–	–	Barreira et al. (2014)
	23	–	–	–	–	–	Villares et al. (2014)
Cantharellus tubaeformis	173	–	–	–	–	–	Teichmann et al. (2007)
Clitocybe odora	–	18.2	ND	117	54.5	190	Vaz et al. (2011)
Coprinus comatus	–	22.8	ND	124	154	301	Vaz et al. (2011)
	–	13.2	ND	ND	31.8	45.0	Stojković et al. (2013)
Craterellus cornucopioides	327	24	155	8.0	ND	187	Barros et al. (2008b)
	–	115	ND	62	17	194	Liu et al. (2012)
	44	–	–	–	–	–	Villares et al. (2014)

(Continued)

Table 2.13 Content of fat-soluble provitamin ergosterol (mg 100 g^{-1} dry matter) and tocopherols (µg 100 g^{-1} dry matter) in selected mushroom species (Continued)

Species	Ergosterol	Tocopherols					Reference
		Alpha-	Beta-	Gamma-	Delta-	Total	
Fistulina hepatica	–	12	173	41	ND	226	Heleno et al. (2010)
	108	–	–	–	–	–	Barreira et al. (2014)
Flammulina velutipes	–	10	30	ND	20	60	Pereira et al. (2012)
Gyromitra esculenta	–	2.2	ND	11.5	99.1	112.8	Leal et al. (2013)
Hydnum repandum	–	ND	ND	51	ND	51	Heleno et al. (2010)
Hygrophorus marzuolus	681	–	–	–	–	–	Vilares et al. (2014)
Laccaria amethystea	–	5.0	109	83	ND	197	Heleno et al. (2010)
	637	212	ND	49	11	272	Liu et al. (2012)
Laccaria laccata	–	22	706	57	19	804	He.eno et al. (2010)
Lactarius deliciosus	84	–	–	–	–	–	Kalogeropoulos et al. (2013)
	55	–	–	–	–	–	Barreira et al. (2014)
	32	–	–	–	–	–	Vilares et al. (2014)
Lactarius sanguifluus	197	–	–	–	–	–	Kalogeropoulos et al. (2013)
Laetiporus sulphureus	–	109	ND	62.1	18.4	190	Petrović et al. (2014)
Lycoperdon echinatum	–	9.4	48.2	73.9	ND	131.5	Grangeia et al. (2011)
Macrolepiota procera	118	–	–	–	–	–	Barreira et al. (2014)
Marasmius oreades	–	6.0	19	130	ND	155	Barros et al. (2008b)
Morchella esculenta	–	2.4	ND	12.4	48.9	63.7	Heleno et al. (2013)
	43	–	–	–	–	–	Barreira et al. (2014)
Pleurotus eryngii	–	6.8	48.2	31.6	ND	86.6	Reis et al. (2014a)

Species							Reference
Russula cyanoxantha	–	10.5	ND	21.9	6.5	38.9	Grangeia et al. (2011)
Russula delica	–	ND	–	10.7	15.3	26.0	Fernandes et al. (2014b)
Russula olivacea	–	11.8	ND	11.8	26.2	49.8	Grangeia et al. (2011)
Russula virescens	–	20.0	21.3	8.0	ND	49.3	Leal et al. (2013)
Suillus granulatus from Portugal	–	17.9	175	ND	102	295	Reis et al. (2014b)
from Serbia	–	6.8	180	13.6	19.8	220	Reis et al. (2014b)
Suillus luteus	–	1270	–	240	ND	1510	Jaworska et al. (2014)
Suillus variegatus	–	20	ND	1440	ND	1460	Pereira et al. (2012)
Tuber aestivum	186	–	–	–	–	–	Villares et al. (2012)
Tuber indicum	137	–	–	–	–	–	Villares et al. (2012)
Tuber melanosporum	190	–	–	–	–	–	Villares et al. (2012)

Cultivated species

Species							Reference
Agaricus bisporus white	–	75	166	ND	ND	241	Barros et al. (2008b)
	474	–	–	–	–	–	Teichmann et al. (2007)
	–	2.6	9.7	17.3	29.8	59.4	Reis et al. (2012)
	352	–	–	–	–	–	Barreira et al. (2014)
	642	–	–	–	–	–	Villares et al. (2014)
brown	–	3.1	Traces	ND	ND	3.2	Jaworska et al. (2015)
	399	–	–	–	–	–	Teichmann et al. (2007)
	–	–	–	–	–	–	Reis et al. (2012)
	77	–	–	–	–	–	Barreira et al. (2014)
	139	3.9	ND	9.6	35.2	48.7	Stojković et al. (2014)
	–	ND	25.3	ND	ND	25.3	Glamočlija et al. (2015)

(Continued)

Table 2.13 Content of fat-soluble provitamin ergosterol (mg 100 g^{-1} dry matter) and tocopherols (µg 100 g^{-1} dry matter) in selected mushroom species (Continued)

Species	Ergosterol	Tocopherols					Reference
		Alpha-	Beta-	Gamma-	Delta-	Total	
Agaricus subrufescens	–	77.8	ND	46.5	ND	124.3	Carneiro et al. (2013)
dried formulation	89	10.1	ND	562	ND	572	Stojković et al. (2014)
Coprinus comatus	–	ND	376	166	46.7	589	Stojković et al. (2013)
Flammulina velutipes	–	0.2	ND	1.3	ND	1.5	Reis et al. (2012)
	189	–	–	–	–	–	Barreira et al. (2014)
	495	–	–	–	–	–	Teichmann et al. (2007)
Lentinula edodes	–	0.5	ND	2.7	2.2	5.4	Reis et al. (2012)
	–	1.0	31.3	ND	ND	32.3	Carneiro et al. (2013)
dried formulation	217	–	–	–	–	–	Barreira et al. (2014)
	364	–	–	–	–	–	Villares et al. (2014)
Pleurotus eryngii	–	0.2	2.0	1.7	0.6	4.5	Reis et al. (2012)
	187	–	–	–	–	–	Barreira et al. (2014)
	419	–	–	–	–	–	Teichmann et al. (2007)
Pleurotus ostreatus	–	0.3	0.9	9.1	3.0	13.3	Reis et al. (2012)
	104	–	–	–	–	–	Barreira et al. (2014)
	331	–	–	–	–	–	Villares et al. (2014)

ND, content below detection limit.

are stored in the hydrophobic core of cytosolic lipid particles and play a role in sterol homeostasis. The ratio of free and esterified ergosterol in the cell is regulated by several factors.

Ultraviolet (UV) light is necessary for the biotransformation of ergosterol to vitamin D_2. Ultraviolet light is classified into three regions: A, B, and C, with wavelengths of 400–315 nm, 315–280 nm (boundary of 320 nm is sometimes used), and less than 280 nm, respectively. Limited data reported higher levels of vitamin D_2 in wild mushrooms exposed to more intensive daylight than in mushrooms cultivated under a roof. Existing research studies (eg, Ko et al., 2008; Krings and Berger, 2014; Roberts et al., 2008) have tested the effect of UV-B irradiation of various intensity for several minutes on vitamin D_2 formation in main cultivated species. The results indicated immediate vitamin D_2 formation up to several mg $100 g^{-1}$ DM. The exposed area of the mushrooms and light intensity were shown to be the important factors because UV light penetrates food materials only a few millimeters, depending on the optical properties of the food. The highest effectiveness of vitamin D_2 formation was observed in mushroom slices, followed by gill side of intact fruit bodies. Similarly, a single layer and even distribution were more efficient dispositions than packaging together in layers. Also, UV-B exposure of *A. bisporus* just prior to harvest (Kristensen et al., 2012) and postharvest pulsed UV light treatment for a few seconds (Kalaras et al., 2012) were very effective in the formation of vitamin D_2. The latter treatment did dot adversely affect quality parameters. Such a point of view has to be considered due to possible formation of free radicals caused by UV irradiation. Another procedure, exposure of *P. ostreatus* freeze-dried powder to UV-B, produced remarkably higher levels of vitamin D_2 during a much shorter application time than in fresh fruit bodies (Wu and Ahn, 2014).

High gamma irradiation doses of 5–10 kGy increased vitamin D_2 content in *A. bisporus* two-fold to three-fold as compared to the level of 224 mg $100 g^{-1}$ in the nonirradiated control (Tsai et al., 2014).

Moreover, vitamin D_4 (22-dihydroergocalciferol) was determined in both wild and cultivated species, particularly following UV irradiation, but with lower contents than ergosterol (Phillips et al., 2012).

It should be considered that vitamin D_2 is degraded in UV-B-treated mushrooms during the storage period. Roberts et al. (2008) observed in brown *A. bisporus* (portobello) apparent first-order kinetics, with the degradation rate constant being $0.025\,h^{-1}$.

The daily dietary allowance of $15\,\mu g$ (600 IU; 1 IU $= 0.025\,\mu g$) of vitamin D is usually recommended in Europe and the United States. There is increasing evidence that populations in areas with limited exposure to sunlight, especially during the winter season, suffer from latent vitamin D deficiency. Among the more endangered groups are vegans and vegetarians, because 7-dehydrocholesterol, the natural precursor of vitamin D_3, occurs only in foods of animal origin. Moreover, melanin in dark-skinned individuals inhibits UV light penetration. Under such conditions the nutritional contribution of edible mushrooms increases and comprises the only nonanimal source of vitamin D. Thus, UV-B-irradiated mushrooms become a valuable and safe source of vitamin D_2 for vitamin D-deficient individuals (Calvo et al., 2013; Simon et al., 2013; Urbain et al., 2011).

A distinct situation was observed in healthy adults. Consuming UV-treated white *A. bisporus* or purified vitamin D_2 with untreated mushrooms provided good absorption of the vitamin and its transformation to the active 25-hydroxyergocalciferol. However, vitamin D status in blood serum was not affected due to the proportional decrease of 25-hydroxycholecalciferol, the active form of vitamin D_3 (Stephensen et al., 2012). Only a limited increase of serum 25-hydroxyergocalciferol was observed in an experimental prediabetic overweight/obese adult cohort deficient in vitamin D following daily consumption of various combinations and doses of untreated/UV-B-treated cooked mushrooms and vitamin D_2/D_3 supplements for 16 weeks (Mehrotra et al., 2014).

Vitamin E is the overall term for sum of four tocopherols (α- to δ-) and four tocotrienols (α- to δ-). The latter components have not yet been reported in mushrooms at detectable levels. Vitamin E is the most efficient, natural, fat-soluble antioxidant and protects cell membrane lipids against their oxidation. Alpha-tocopherol is biologically the most efficient, and the activity of further tocopherols decreases in the following order: β- > γ- > δ-. The recent recommended daily intake is approximately 20–30 mg. From this point of view (data for total tocopherols presented in Table 2.13), usually tens to hundreds of μg per 100 g DM (ie, 10-times less in FM) classify mushrooms as food items with low vitamin E content and nutritional contribution.

Moreover, there is wide variability among the individual tocopherols contents and compositions in various species (Table 2.12), even though most of the data originate from the same laboratory, and thus demands further research.

2.6.2 Water-Soluble Vitamins

The informative data on six water-soluble vitamins are presented in Table 2.14. Nevertheless, the range of the analyzed species has been limited and future data may contribute some corrective information. Unfortunately, information on expectable losses during preservation, storage, and cooking treatments and on bioavailability has been lacking.

Literature data on ascorbic acid (vitamin C) mostly report contents between 100 and 400 mg 100 g^{-1} DM (ie, approximately 10–40 mg 100 g^{-1}) in fresh mushrooms. This is comparable with potatoes, but lower than in most vegetables. Losses during storage, under usual preservation methods, and during cooking treatments have to be supposed.

Within the vitamins of the B group, the contents of niacin (vitamin PP or B$_3$) are similar to usual levels in common vegetables; those of riboflavin (vitamin B$_2$) are lower. As compared to other vegetables, mushrooms are poorer in thiamin (vitamin B$_1$)

Table 2.14 Mean contents of water-soluble vitamins (mg or µg 100 g^{-1} dry matter)

Species	Ascorbic acid Vitamin C (mg)	Thiamine Vitamin B$_1$ (mg)	Riboflavin Vitamin B$_2$ (mg)	Folates Vitamin B$_9$ (mg)	Cyanocobalamin Vitamin B$_{12}$ (µg)	Niacin/niacinamid Vitamir PP or B$_3$ (mg)	Reference
Wild-growing species							
Cantharellus cibarius	–	–	–	–	1.46	–	Watanabe et al. (2012)
Craterellus cornucopioides	–	–	–	–	2.19	–	Watanabe et al. (2012)
Suillus luteus	25.6	1.1	3.9	–	–	35	Jaworska et al. (2014)
Cultivated species							
Agaricus bisporus							
white	19	0.6	5.1	0.45	0.8	43	Mattila et al. (2001)
	–	–	–	–	1.01	–	Furlani and Godoy (2007)
	–	1.05	3.6	–	–	–	Bernaś and Jaworska (2015)
	7.2	0.9	5.5	–	–	24	Jaworska et al. (2015)
brown	21	0.6	4.2	0.59	0.6	53	Mattila et al. (2001)
Lentinula edodes	25	0.6	1.8	0.3	0.8	31	Mattila et al. (2001)
	–	–	–	0.66	–	–	Furlari and Godoy (2007)
	–	–	–	–	5.6 ± 3.9[a]	–	Bito et al. (2014)
	–	–	–	–	4.2 ± 2.4[b]	–	Bito et al. (2014)
Pleurotus ostreatus	20	0.9	2.5	0.64	0.6	65	Mattila et al. (2001)
	–	–	–	0.79	–	–	Furlani and Godoy (2007)
	28–35	1.9–2.0	1.8–5.1	0.3–0.7	–	30–65	Khan and Tania (2012)
	5.9	0.3	1.6	–	–	10	Jaworska et al. (2015)

[a]Dried Donko-type fruit bodies with closed caps.
[b]Dried Koushin-type fruit bodies with open caps.

and folate (vitamin B_9). Information on pyridoxine (vitamin B_6) has been very limited. Contents of 0.95 and 0.07 mg $100 g^{-1}$ FM were determined in cultivated white *A. bisporus* and *P. ostreatus*, respectively. Pyridoxine highly prevailed in *A. bisporus*, whereas the same levels of pyridoxine and pyridoxamine were found in *P. ostreatus*. Pyridoxal was a minor component in both species (Jaworska et al., 2015). Losses of thiamin and riboflavin at levels of 14–65% and 3–13% of the initial contents in fresh *P. ostreatus*, respectively, were reported after 12 months of storage at −25°C. The differences were caused by various preprocessing of fruit bodies prior to storage (Jaworska and Bernaś, 2009). Significant losses in thiamin, riboflavin, niacin, and pyridoxine were observed during blanching, stewing, and subsequent storage of *A. bisporus* and *P. ostreatus* (Jaworska et al., 2015).

Vitamin B_{12}, usually restricted to cyanocobalamin, is synthesized by certain bacteria present in animal organisms. Animal foods are thus considered to be the major dietary sources. Typical levels of vitamin B_{12} in foods range from low $\mu g\ kg^{-1}$ for cheese and fish to hundreds of $\mu g\ kg^{-1}$ for liver. Vegans and vegetarians thus have an elevated risk for vitamin B_{12} deficiency. As shown in Table 2.14, considerable contents were observed in *C. cibarius* and *C. cornucopioides*, and even higher contents were found in *L. edodes*. The determined compound was identified as authentic vitamin B_{12} (cyanocobalamin), not pseudovitamin B_{12}, occurring in some edible algae, which is inactive in humans (Watanabe et al., 2012). Vitamin B_{12} production in mushrooms independent of bacteria is controversial. It has been suggested that cyanocobalamin is produced by microorganisms living in substrate and on the surface of mushrooms. Higher vitamin B_{12} contents were detected in the outer peel of *A. bisporus* than in other parts of the mushroom (Koyyalamudi et al., 2009). Bito et al. (2014) concluded that vitamin B_{12} found in dried *L. edodes* (Table 2.14) had been derived from bed logs used for the cultivation.

Very high losses of approximately 75% of the initial content of ascorbic acid, thiamine, and riboflavine caused usual stewing of *Suillus luteus*. Cold storage of the stewed mushrooms led to further losses (Jaworska et al., 2014).

Overall, mushrooms have been often popularized as a good source of vitamins (or provitamins). However, the information should be limited to ergosterol, vitamin D_2, and vitamin B_{12}.

REFERENCES

Akata, I., Ergönül, B., Kalyoncu, F., 2012. Chemical composition and antioxidant activities of 16 wild edible mushroom species grown in Anatolia. Int. J. Pharmacol. 8, 134–138.

Akata, I., Ergönül, P.G., Ergönül, B., Kalyoncu, F., 2013. Determination of fatty acid contents of five wild edible mushroom species collected from Anatolia. J. Pure Appl. Microbiol. 7, 3143–3147.

Akyüz, M., Kirbağ, S., 2010. Nutritive value of wild edible and cultured mushrooms. Turk. J. Biol. 34, 97–102.

Ayaz, F.A., Torun, H., Özel, A., Col, M., Duran, C., Sesli, E., et al., 2011a. Nutritional value of some wild edible mushrooms from the Black Sea region (Turkey). Turk. J. Biol. 36, 385–393.

Ayaz, F.A., Chuang, L.T., Torun, H., Colak, A., Sesli, E., Presley, J., et al., 2011b. Fatty acid and amino acid composition of selected wild-edible mushrooms consumed in Turkey. Int. J. Food Sci. Nutr. 62, 328–335.

Barreira, J.C.M., Ferreira, I.C.F.R., Oliveira, M.B.P.P., 2012. Triacylglycerol profile as a chemical fingerprint of mushroom species: evaluation by principal component and linear discriminant analyses. J. Agric. Food Chem. 60, 10592–10599.

Barreira, J.C.M., Oliveira, M.B.P.P., Ferreira, I.C.F.R., 2014. Development of a novel methodology for the analysis of ergosterol in mushrooms. Food Anal. Methods 7, 217–223.

Barros, L., Baptista, P., Correira, D.M., Casa, S., Oliveira, B., Ferreira, I.C.F.R., 2007a. Fatty acid and sugar compositions, and nutritional value of five wild edible mushrooms from Northeast Portugal. Food Chem. 105, 140–145.

Barros, L., Baptista, P., Correia, D.M., Morais, J.S., Ferreira, I.C.F.R., 2007b. Effects of conservation treatment and cooking on the chemical composition and antioxidant activity of Portuguese wild edible mushrooms. J. Agric. Food Chem. 55, 4781–4788.

Barros, L., Ventuizini, B.A., Baptista, P., Estevinho, L.M., Ferreira, I.C.F.R., 2008a. Chemical composition and biological properties of Portuguese wild mushrooms: a comprehensive study. J. Agric. Food Chem. 56, 3856–3862.

Barros, L., Cruz, T., Baptista, P., Estevinho, L.M., Ferreira, I.C.F.R., 2008b. Wild and commercial mushrooms as source of nutrients and nutraceuticals. Food Chem. Toxicol. 46, 2742–2747.

Barros, L., Correia, D.M., Ferreira, I.C.F.R., Baptista, P., Santos-Buelga, C., 2008c. Optimization of the determination of tocopherols in *Agaricus* sp. edible mushrooms by a normal phase liquid chromatographic method. Food Chem. 110, 1046–1050.

Bauer Petrovska, B., 2001. Protein fraction in edible Macedonian mushrooms. Eur. Food Res. Technol. 212, 469–472.

Bauer Petrovska, B., 2002. Infrared analysis of Macedonian mushroom dietary fibre. Nahrung/Food 46, 238–239.

Beluhan, S., Ranogajec, A., 2011. Chemical composition and non-volatile components of Croatian wild edible mushrooms. Food Chem. 124, 1076–1082.

Bernaś, E., Jaworska, G., 2015. Effect of microwave blanching on the quality of frozen *Agaricus bisporus*. Food Sci. Technol. Int. 21, 245–255.

Bito, T., Teng, F., Ohishi, N., Takenaka, S., Miyamoto, E., Sakuno, E., et al., 2014. Characterization of vitamin B_{12} compounds in the fruiting bodies of shiitake mushroom (*Lentinula edodes*) and bed logs after fruiting of the mushroom. Mycoscience 55, 462–468.

Braaksma, A., Schaap, D.J., 1996. Protein analysis of the common mushroom *Agaricus bisporus*. Postharvest Biol. Technol. 7, 119–127.

Çağlarirmak, N., 2009. Determination of nutrients and volatile constituents of *Agaricus bisporus* (brown) at different stages. J. Sci. Food Agric. 89, 634–638.

Calvo, M.S., Babu, U.S., Garthoff, L.H., Woods, T.O., Dreher, M., Hill, G., et al., 2013. Vitamin D_2 from light-exposed edible mushrooms is safe, bioavailable and effectively supports bone growth in rats. Osteoporos. Int. 24, 197–207.

Carneiro, A.A.J., Ferreira, I.C.F.R., Dueñas, M., Barros, L., da Silva, R., Gomes, E., et al., 2013. Chemical composition and antioxidant activity of dried powder formulations of *Agaricus blazei* and *Lentinus edodes*. Food Chem. 138, 2168–2173.

Carvalho, L.M., Carvalho, F., de Lourdes Bastos, M., Baptista, P., Moreira, N., Monforte, A.R., et al., 2014. Non-targeted and targeted analysis of wild toxic and edible mushrooms using gas chromatography-ion trap mass spectrometry. Talanta 118, 292–303.

Chen, W., Li, W., Yang, Y., Yu, H., Zhou, S., Feng, J., et al., 2015. Analysis and evaluation of tasty components in the pileus and stipe of *Lentinula edodes* at different growth stages. J. Agric. Food Chem. 63, 795–801.

Cheung, P.C.-K., 1996. Dietary fiber content and composition of some cultivated edible mushroom fruiting bodies and mycelia. J. Agric. Food Chem. 44, 468–471.

Cheung, P.C.-K., 2013. Mini-review on edible mushrooms as source of dietary fiber: preparation and health benefits. Food Sci. Hum. Wellness 2, 136–166.

Cohen, N., Cohen, J., Asatiani, M.D., Varshney, V.K., Yu, H.T., Yang, Y.C., et al., 2014. Chemical composition and nutritional and medicinal value of fruit bodies and submerged cultured mycelia of culinary-medicinal higher Basidiomycetes mushrooms. Int. J. Med. Mushrooms 16, 273–291.

Cui, F., Li, Y., Yang, Y., Sun, W., Wu, D., Ping, L., 2014. Changes in chemical components and cytotoxicity at different maturity stages of *Pleurotus eryngii* fruiting body. J. Agric. Food Chem. 62, 12631–12640.

Dembitsky, V.M., Terent'ev, A.O., Levitsky, D.O., 2010. Amino and fatty acids of wild edible mushrooms of the genus *Boletus*. Rec. Nat. Prod. 4, 218–223.

Díez, A.A., Alvarez, A., 2001. Compositional and nutritional studies on two wild edible mushrooms from northwest Spain. Food Chem. 75, 417–422.

Dikeman, C.L., Bauer, L.L., Flickinger, E.A., Fahey Jr., G.C., 2005. Effects of stage of maturity and cooking on the chemical composition of select mushroom varieties. J. Agric. Food Chem. 53, 1130–1138.

Doğan, H.H., 2013. Evaluation of phenolic compounds, antioxidant activities and fatty acid composition of *Amanita ovoidea* (Bull.) Link. in Turkey. J. Food Compost. Anal. 31, 87–93.

Doğan, H.H., Akbaş, G., 2013. Biological activity and fatty acid composition of Caesar's mushroom. Pharm. Biol. 51, 863–871.

Dundar, A., Faruk Yesil, O., Acay, H., Okumus, V., Ozdemir, S., Yildiz, A., 2012. Antioxidant properties, chemical composition and nutritional value of *Terfezia boudieri* (Chatin) from Turkey. Food Sci. Technol. Int. 18, 317–328.

Ergönül, P.G., Kalyoncu, F., Ergönül, B., 2013. Fatty acid compositions of six wild edible mushroom species. Sci. World J.

Fernandes, Â., Barros, L., Barreira, J.C.M., Antonio, A.L., Oliveira, M.B.P.P., Martins, A., et al., 2013a. Effect of different processing technologies on chemical and antioxidant parameters of *Macrolepiota procera* wild mushroom. LWT—Food Sci. Technol. 54, 493–499.

Fernandes, Â., Antonio, A.L., Barreira, J.C.M., Botelho, M.L., Oliveira, M.B.P.P., Martins, A., et al., 2013b. Effect of gamma irradiation on the chemical composition and antioxidant activity of *Lactarius deliciosus* L. wild edible mushroom. Food Bioprocess Technol. 6, 2895–2903.

Fernandes, Â., Barreira, J.C.M., Antonio, A.L., Santos, P.M.P., Martins, A., Oliveira, M.B.P.P., et al., 2013c. Study of chemical changes and antioxidant activity variation induced by gamma-irradiation on wild mushrooms: comparative study through principal component analysis. Food Res. Int. 54, 18–25.

Fernandes, Â., Barreira, J.C.M., Antonio, A.L., Oliveira, M.B.P.P., Martins, A., Ferreira, I.C.F.R., 2014a. Effect of gamma irradiation on chemical composition and antioxidant potential of processed samples of the wild mushroom *Macrolepiota procera*. Food Chem. 149, 91–98.

Fernandes, Â., Barreira, J.C.M., Antonio, A.L., Oliveira, M.B.P.P., Martins, A., Ferreira, I.C.F.R., 2014b. Feasibility of electron-beam irradiation to preserve wild dried mushrooms: effects on chemical composition and antioxidant activity. Innov. Food Sci. Eng. Technol. 22, 158–166.

Fernandes, Â., Barros, L., Martins, A., Herbert, P., Ferreira, I.C.F.R., 2015a. Nutritional characterisation of *Pleurotus ostreatus* (Jacq. ex Fr.) P. Kumm. produced using paper scrabs as substrate. Food Chem. 169, 396–400.

Fernandes, Â., Barreira, J.C.M., Antonio, A.L., Morales, P., Férnandez-Ruiz, V., Martins, A., et al., 2015b. Exquisite wild mushrooms as a source of dietary fiber: analysis in electron-beam irradiated samples. LWT—Food Sci. Technol. 60, 855–859.

Fernandes, Â., Barreira, J.C.M., Antonio, A.L., Rafalski, A., Oliveira, M.B.P.P., Martins, A., et al., 2015c. How does electron beam irradiation dose affect the chemical and antioxidant profiles of wild dried *Amanita* mushrooms? Food Chem. 182, 309–315.

Florczak, J., Karmańska, A., Wędzisz, A., 2004. [A comparison of chemical composition of selected wild-growing mushrooms]. Bromatologia i Chemiczna Toksykologia 37, 365–370. (in Polish).

Furlani, R.P.Z., Godoy, H.T., 2007. Contents of folates in edible mushrooms commercialised in the city of Campinas, São Paulo, Brazil. Ciéncia e Tecnologia de Alimentos 27, 278–280.

Glamočlija, J., Stojković, D., Nikolić, M., Ćirić, A., Reis, F.S., Barros, L., et al., 2015. A comparative study on edible *Agaricus* mushrooms as functional foods. Food Funct. 6, 1900–1910.

Gogavekar, S.S., Rokade, S.A., Ranveer, R.C., Ghosh, J.S., Kalyani, D.C., Sahoo, A.K., 2014. Important nutritional constituents, flavour components, antioxidant and antibacterial properties of *Pleurotus sajor-caju*. J. Food Sci. Technol. 51, 1483–1491.

Grangeia, C., Heleno, S.A., Barros, L., Martins, A., Ferreira, I.C.F.R., 2011. Effect of trophism on nutritional and nutraceutical potential of wild edible mushrooms. Food Res. Int. 44, 1029–1035.

Györfi, J., Geösel, A., Vetter, J., 2010. Mineral composition of different strains of edible medicinal mushroom *Agaricus subrufescens* Peck. J. Med. Food 13, 1510–1514.

Heleno, S.A., Barros, L., Sousa, M.J., Martins, A., Ferreira, I.C.F.R., 2009. Study and characterization of selected nutrients in wild mushrooms from Portugal by gas chromatography and high performance liquid chromatography. Microchem. J. 93, 195–199.

Heleno, S.A., Barros, L., Sousa, M.J., Martins, A., Ferreira, I.C.F.R., 2010. Tocopherols composition of Portuguese wild mushrooms with antioxidant capacity. Food Chem. 119, 1443–1450.

Heleno, S.A., Barros, L., Sousa, M.J., Martins, A., Santos-Buelga, C., Ferreira, I.C.F.R., 2011. Targeted metabolites analysis in wild *Boletus* species. LWT—Food Sci. Technol. 44, 1343–1348.

Heleno, S.A., Stojković, D., Barros, L., Glamočlija, J., Soković, M., Martins, A., et al., 2013. A comparative study of chemical composition, antioxidant and antimicrobial properties of *Morchella esculenta* (L.) Pers. from Portugal and Serbia. Food Res. Int. 51, 236–243.

Heleno, S.A., Barros, L., Martins, A., Queiroz, M.J.R.P., Morales, P., Fernández-Ruiz, V., et al., 2015. Chemical composition, antioxidant activity and bioaccessibility studies in phenolic extracts of two *Hericium* wild edible species. LWT—Food Sci. Technol. 63, 475–481.

Jaworska, G., Bernaś, E., 2009. Qualitative changes in *Pleurotus ostreatus* (Jacq.:Fr.) Kumm. mushrooms resulting from different methods of preliminary processing and periods of frozen storage. J. Sci. Food Agric. 89, 1066–1075.

Jaworska, G., Bernaś, E., 2012. Amino acid content of frozen *Agaricus bisporus* and *Boletus edulis* mushrooms: effects of pretreatments. Int. J. Food Prop. 16, 139–153.

Jaworska, G., Bernaś, E., Mickowska, B., 2011. Effect of production process on the amino acid content of frozen and canned *Pleurotus ostreatus* mushrooms. Food Chem. 125, 936–943.

Jaworska, G., Bernaś, E., Biernacka, A., 2012. Effect of pretreatments and storage on the amino acid content of canned mushrooms. J. Food Process. Preserv. 36, 242–251.

Jaworska, G., Pogoń, K., Bernaś, E., Skrzypczak, A., Kapusta, I., 2014. Vitamins, phenolics and antioxidant activity of culinary prepared *Suillus luteus* (L.) Roussel mushroom. LWT—Food Sci. Technol. 59, 701–706.

Jaworska, G., Pogoń, K., Bernaś, E., Duda-Chodak, A., 2015. Nutraceuticals and antioxidant activity of prepared for consumption commercial mushrooms *Agaricus bisporus* and *Pleurotus ostreatus*. J. Food Qual. 38, 111–122.

Kalaras, M.D., Beelman, R.B., Elias, R.J., 2012. Effects of postharvest pulsed UV light treatment of white button mushrooms (*Agaricus bisporus*) on vitamin D_2 content and quality attributes. J. Agric. Food Chem. 60, 220–225.

Kalogeropoulos, N., Yanni, A.E., Koutrotsios, G., Aloupi, M., 2013. Bioactive microconstituents and antioxidant properties of wild edible mushrooms from the island of Lesvos, Greece. Food Chem. Toxicol. 55, 378–385.

Karliński, L., Ravnskov, S., Kieliszewska-Rokicka, B., Larsen, J., 2007. Fatty acid composition of various ectomycorrhizal fungi and ectomycorrhizas of Norway spruce. Soil Biol. Biochem. 39, 854–866.

Kavishree, S., Hemavathy, J., Lokesh, B.R., Shashirekha, M.N., & Rajarathnam, S., 2008. Fat and fatty acids of Indian edible mushrooms. Food Chem. 106, 597–602.

Khan, A., Tania, M., 2012. Nutritional and medicinal importance of *Pleurotus* mushrooms: an overview. Food Rev. Int. 28, 313–329.

Kim, M.Y., Chung, I.M., Lee, S.J., Ahn, J.K., Kim, E.H., Kim, M.J., et al., 2009. Comparison of free amino acid, carbohydrates concentrations in Korean edible and medicinal mushrooms. Food Chem. 113, 386–393.

Ko, J.A., Lee, B.H., Lee, J.S., Park, H.J., 2008. Effect of UV-B exposure on the concentration of vitamin D_2 in sliced shiitake mushroom (*Lentinus edodes*) and white button mushroom (*Agaricus bisporus*). J. Agric. Food Chem. 56, 3671–3674.

Kovács, D., Vetter, J., 2015. Chemical composition of the mushroom *Laetiporus sulphureus* (Bull.) Murill. Acta Alimentaria 44, 104–110.

Koyyalamudi, S.R., Jeong, S.C., Cho, K.Y., Pang, G., 2009. Vitamin B_{12} is the active corrinoid produced in cultivated white button mushrooms (*Agaricus bisporus*). J. Agric. Food Chem. 57, 6327–6333.

Krings, U., Berger, R.G., 2014. Dynamics of sterols and fatty acids during UV-B treatment of oyster mushroom. Food Chem. 149, 10–14.

Kristensen, H.L., Rosenqvist, E., Jakobsen, J., 2012. Increase of vitamin D_2 by UV-B exposure during the growth phase of white button mushroom (*Agaricus bisporus*). Food Nutr. Res.http://dx.doi.org/10.3402/fnr. v56i0.7114.

Kruzselyi, D., Vetter, J., 2014. Complex chemical evaluation of the summer truffle (*Tuber aestivum* Vittadini) fruit bodies. J. Appl. Bot. Food Qual. 87, 291–295.

Leal, A.R., Barros, L., Barreira, J.C.M., Sousa, M.J., Martins, A., Santos-Buelga, C., et al., 2013. Portuguese wild mushrooms at the "pharma-nutrition" interface: nutritional characterization and antioxidant properties. Food Res. Int. 50, 1–9.

Lee, Y.L., Jian, S.Y., Mau, J.L., 2009. Composition and non-volatile taste components of *Hypsizigus marmoreus*. LWT—Food Sci. Technol. 42, 594–598.

Li, W., Gu, Z., Zhou, S., Liu, Y., Zhang, J., 2014. Non-volatile taste components of several cultivated mushrooms. Food Chem. 143, 427–431.

Li, X., Feng, T., Zhou, F., Zhou, S., Liu, Y., Li, W., et al., 2015. Effects of drying methods on the tasty compounds of *Pleurotus eryngii*. Food Chem. 166, 358–364.

Liu, Y.T., Sun, J., Luo, S.Q., Su, Y.J., Xu, R.R., Yang, Y.J., 2012. Chemical composition of five wild edible mushrooms collected from Southwest China and their antihyperglycemic and antioxidant activity. Food Chem. Toxicol. 50, 1238–1244.

Liu, Y., Huang, F., Yang, H., Ibrahim, S.A., Wang, Y., Huang, W., 2014. Effects of preservation methods on amino acids and 5′-nucleotides of *Agaricus bisporus* mushrooms. Food Chem. 149, 221–225.

Manzi, P., Gambelli, L., Marconi, S., Vivanti, V., Pizzoferrato, L., 1999. Nutrients in edible mushrooms: an inter-species comparative study. Food Chem. 65, 477–482.

Manzi, P., Aguzzi, A., Pizzoferrato, L., 2001. Nutritional value of mushrooms widely consumed in Italy. Food Chem. 73, 321–325.

Manzi, P., Marconi, S., Aguzzi, A., Pizzoferrato, L., 2004. Commercial mushrooms: nutritional quality and effect of cooking. Food Chem. 84, 201–206.

Mattila, P., Könkö, K., Pihlava, J.-M., Astola, J., Vahteristo, L., Hietaniemi, V., et al., 2001. Contents of vitamins, mineral elements, and some phenolic compounds in cultivated mushrooms. J. Agric. Food Chem. 49, 2343–2348.

Mattila, P., Salo-Vaananen, P., Könkö, K., Aro, H., Jalava, T., 2002. Basic composition and amino acid contents of mushrooms cultivated in Finland. J. Agric. Food Chem. 50, 6419–6422.

Mehrotra, A., Calvo, M.S., Beelman, R.B., Levy, E., Siuty, J., Kalaras, M.D., et al., 2014. Bioavailability of vitamin D-2 from enriched mushrooms in prediabetic adults: a randomized controlled trial. Eur. J. Clin. Nutr. 68, 1154–1160.

Murray, I.A., Coupland, K., Smith, J.A., Ansell, I.D., Long, R.G., 2000. Intestinal trehalase activity in a UK population: establishing a normal range and the effect of disease. Br. J. Nutr. 83, 241–245.

Nasr, M., Malloch, D.W., Arp, P.A., 2012. Quantifying Hg within ectomycorrhizal fruiting bodies, from emergence to senescence. Fungal Biol. 116, 1163–1177.

Nile, S.H., Park, S.W., 2014. Total, soluble, and insoluble dietary fibre contents of wild growing edible mushrooms. Czech J. Food Sci. 32, 302–307.

Nitschke, J., Altenbach, H.-J., Malolepszy, T., Möllcken, H., 2011. A new method for the quantification of chitin and chitosan in edible mushrooms. Carbohydr. Res. 346, 1307–1310.

Obodai, M., Ferreira, I.C.F.R., Fernandes, A., Barro, L., Narh Mensah, D.L., Dzomeku, M., et al., 2014. Evaluation of the chemical and antioxidant properties of wild and cultivated mushrooms of Ghana. Molecules 19, 19532–19548.

Ouzouni, P.K., Riganakos, K.A., 2007. Nutritional value and metal content of Greek wild edible fungi. Acta Alimentaria 36, 99–110.

Ouzuni, P.K., Petridis, D., Koller, W.D., Riganakos, K.A., 2009. Nutritional value and metal content of wild edible mushrooms collected from West Macedonia and Epirus, Greece. Food Chem. 115, 1575–1580.

Öztürk, M., Tel, G., Öztürk, F.A., Duru, M.E., 2014. The cooking effect on two edible mushrooms in Anatolia: fatty acid composition, total bioactive compounds, antioxidant and anticholinesterase activity. Rec. Nat. Prod. 8, 189–194.

Pedneault, K., Angers, P., Gosselin, A., Tweddel, R.J., 2006. Fatty acid composition of lipids from mushrooms belonging to the family Boletaceae. Mycol. Res. 110, 1179–1183.

Pedneault, K., Angers, P., Gosselin, A., Tweddel, R.J., 2008. Fatty acid profiles of polar and neutral lipids of ten species of higher basidiomycetes indigenous to eastern Canada. Mycol. Res. 112, 1428–1434.

Pei, F., Shi, Y., Gao, X., Wu, F., Mariga, A.M., Yang, W., et al., 2014. Changes in non-volatile taste components of button mushroom (*Agaricus bisporus*) during different stages of freeze drying and freeze drying combined with microwave vacuum drying. Food Chem. 165, 547–554.

Pereira, E., Barros, L., Martins, A., Ferreira, I.C.F.R., 2012. Towards chemical and nutritional inventory of Portuguese wild edible mushrooms in different habitats. Food Chem. 130, 394–403.

Petrović, J., Stojković, D., Reis, F.S., Barros, L., Glamočlija, J., Ćirić, A., et al., 2014. Study on chemical, bioactive and food preserving properties of *Laetiporus sulphureus* (Bull.: Fr.) Murr. Food Funct. 5, 1441–1451.

Phillips, K.M., Horst, R.L., Koszewski, N.J., Simon, R.R., 2012. Vitamin D_4 in mushrooms. Plos One. http://dx.doi.org/10.1371/journal.pone.0040702.

Quinche, J.-P., 1997. [Phosphorus and heavy metals in some species of fungi.]. Rev. Suisse Agric. 29, 151–156. (in French).

Reis, F.S., Barros, L., Martins, A., Ferreira, I.C.F.R., 2012. Chemical composition and nutritional value of the most widely appreciated cultivated mushrooms: an inter-species comparative study. Food Chem. Toxicol. 50, 191–197.

Reis, F.S., Barros, L., Sousa, M.J., Martins, A., Ferreira, I.C.F.R., 2014a. Analytical methods applied to the chemical characterization and antioxidant properties of three wild edible mushroom species from Northeastern Portugal. Food Anal. Methods 7, 645–652.

Reis, F.S., Stojković, D., Barros, L., Glamočlija, J., Ćirić, A., Soković, M., et al., 2014b. Can *Suillus granulatus* (L.) Roussel be classified as a functional food? Food and Funct. 5, 2861–2869.

Ribeiro, B., de Pinho, P.G., Andrade, P.B., Baptista, P., Valentão, P., 2009. Fatty acid composition of wild edible mushroom species: a comparative study. Microchem. J. 93, 29–35.

Roberts, J.S., Teichert, A., McHugh, T.H., 2008. Vitamin D_2 formation from post-harvest UV-B treatment of mushrooms (*Agaricus bisporus*) and retention during storage. J. Agric. Food Chem. 56, 4541–4544.

Rudawska, M., Leski, T., 2005. Macro- and microelement contents in fruiting bodies of wild mushrooms from the Notecka forest in west-central Poland. Food Chem. 92, 499–506.

Shao, S., Hernandez, M., Kramer, J.K.G., Rinker, D.L., Tsao, R., 2010. Ergosterol profiles, fatty acid composition, and antioxidant activities of button mushrooms as affected by tissue part and developmental stage. J. Agric. Food Chem. 58, 11616–11625.

Simon, R.R., Borzelleca, J.F., DeLuca, H.F., Weaver, C.M., 2013. Safety assessment of the post-harvest treatment of button mushrooms (*Agaricus bisporus*) using ultraviolet light. Food Chem. Toxicol. 56, 278–289.

Stephensen, C.B., Zerofsky, M., Burnett, D.J., Lin, Y.P., Hammock, B.D., Hall, L.M., et al., 2012. Ergocalciferol from mushrooms or supplements consumed with a standard meal increases 25-hydroxycalciferol but decreases 25-hydroxycholecalciferol in the serum of healthy adults. J. Nutr. 142, 1246–1252.

Stojković, D., Reis, F.S., Barros, L., Glamočlija, J., Ćirić, A., van Griensven, L.J.L.D., et al., 2013. Nutrients and non-nutrients composition and bioactivity of wild and cultivated *Coprinus comatus* (O.F.Müll.) Pers. Food Chem. Toxicol. 59, 289–296.

Stojković, D., Reis, F.S., Glamočlija, J., Ćirić, A., Barros, L., van Griensven, L.J.L.D., et al., 2014. Cultivated strains of *Agaricus bisporus* and *A. brasiliensis*: chemical characterization and evaluation of antioxidant and antimicrobial properties for the final healthy product—natural preservatives in yoghurt. Food Funct. 5, 1602–1612.

Surinrut, P., Julshamn, K., Njaa, L.E.R., 1987. Protein, amino acids and some major and trace elements in Thai and Norwegian mushrooms. Plant. Foods Hum. Nutr. 37, 117–125.

Teichmann, A., Dutta, P.C., Staffas, A., Jägerstad, M., 2007. Sterol and vitamin D_2 concentration in cultivated and wild grown mushrooms: effects of UV irradiation. LWT—Food Sci. Technol. 40, 815–822.

Tsai, S.Y., Tsai, H.L., Mau, J.L., 2008. Non-volatile taste components of *Agaricus blazei*, *Agrocybe cylindracea* and *Boletus edulis*. Food Chem. 107, 977–983.

Tsai, S.Y., Mau, J.L., Huang, S.J., 2014. Enhancement of antioxidant properties and increase of content of vitamin D_2 and non-volatile components in fresh button mushroom, *Agaricus bisporus* (higher Basidiomycetes) by gamma-irradiation. Int. J. Med. Mushrooms 16, 137–147.

Ulziijargal, E., Mau, J.L., 2011. Nutrient compositions of culinary-medicinal mushroom fruiting bodies and mycelia. Int. J. Med. Mushrooms 13, 343–349.

Urbain, P., Singler, F., Ihorst, G., Biesalski, H.-K., Bertz, H., 2011. Bioavailability of vitamin D_2 from UV-B-irradiated button mushrooms in healthy adults in serum 25-hydroxyvitamin D: a randomized trial. Eur. J. Clin. Nutr. 65, 965–971.

Vaz, J.A., Barros, L., Martins, A., Santos-Buelga, C., Vasconselos, M.H., Ferreira, I.C.F.R., 2011. Chemical composition of wild edible mushrooms and antioxidant properties of their water soluble polysaccharidic and ethanolic fractions. Food Chem. 126, 610–616.

Vetter, J., 1993a. [Chemical composition of eight edible mushrooms]. Zeitschrift für Lebensmittel Untersuchung und –Forschung 196, 224–227. (in German).

Vetter, J., 1993b. [Amino acid composition of edible mushrooms of genera *Russula* and *Agaricus*]. Zeitschrift für Lebensmittel Untersuchung und –Forschung 197, 381–384. (in German).

Vetter, J., 2007. Chitin content of cultivated mushrooms *Agaricus bisporus*, *Pleurotus ostreatus* and *Lentinula edodes*. Food Chem. 102, 6–9.

Vetter, J., Rimóczi, I., 1993. [Crude, digestible and indigestible protein in fruiting bodies of *Pleurotus ostreatus*]. Zeitschrift für Lebensmittel Untersuchung und –Forschung 197, 427–428. (in German).

Villares, A., García-Lafuente, A., Guillamón, E., Ramos, Á., 2012. Identification and quantification of ergosterol and phenolic compounds occurring in *Tuber* spp. truffles. J. Food Compost. Anal. 26, 177–182.

Villares, A., Mateo-Vivaracho, L., García-Lafuente, A., Guillamón, E., 2014. Storage temperature and UV-irradiation influence on the ergosterol content in edible mushrooms. Food Chem. 147, 252–256.

Wang, S., Marcone, M.F., 2011. The biochemistry and biological properties of the world's most expensive underground edible mushroom: truffles. Food Res. Int. 44, 2567–2581.

Wang, X.M., Yhang, J., Wu, L.H., Yhao, Z.L., Li, T., Li, J.Q., et al., 2014. A mini-review of chemical composition and nutritional value of edible wild-grown mushroom from China. Food Chem. 151, 279–285.

Watanabe, F., Schwarz, J., Takenaka, S., Miyamoto, E., Ohishi, N., Nelle, E., et al., 2012. Characterization of vitamin B_{12} compounds in the wild edible mushroom black trumpet (*Craterellus cornucopioides*) and golden chanterelle (*Cantharellus cibarius*). J. Nutr. Sci. Vitaminol. (Tokyo) 58, 438–441.

Wu, W.J., Ahn, B.Y., 2014. Statistical optimization of ultraviolet conditions for vitamin D-2 synthesis in oyster mushrooms (*Pleurotus ostreatus*) using response surface methodology. Plos One 9. http://dx.doi.org/10.1371/journal.pone.0095359.

Wu, F., Tang, J., Pei, F., Wang, S., Chen, G., Hu, Q., et al., 2015. The influence of four drying methods on nonvolatile taste components of White *Hypsizygus marmoreus*. Eur. Food Res. Technol. 240, 823–830.

Yang, J.H., Lin, H.C., Mau, J.L., 2001. Non-volatile taste components of several commercial mushrooms. Food Chem. 72, 465–471.

Zhou, S., Tang, Q.J., Zhang, Z., Li, C., Cao, H., Yang, Y., et al., 2015. Nutritional composition of three domesticated culinary-medicinal mushrooms: *oudemansiella submucida*, *Lentinus squarrosulus*, and *Tremella aurantialba*. Int. J. Med. Mushrooms 17, 43–49.

CHAPTER 3

Minor Constituents

Contents

Edible mushrooms contain low concentrations of numerous compounds of various chemical structures and, thus, diversified biological properties. The constituents affecting sensory properties together with compounds showing both favorable and adverse effects (eg, various trace elements) are presented in this chapter. The constituents with obvious health-stimulating or deleterious effects are described in the following chapters.

3.1 TASTE AND FLAVOR COMPONENTS

The specific taste of edible mushrooms originates from various combinations of water-soluble substances such as free amino acids (Table 3.1), 5′-nucleotides (Table 3.2), sugars and polyols, and

Table 3.1 Content of free amino acids (mg g^{-1} dry matter) affecting the taste of fresh mushrooms

Species	MSG-like	Sweet	Bitter	Tasteless	Total	Reference
Wild-growing species						
Agaricus campestris	35.0	16.9	5.77	5.27	62.9	Beluhan and Ranogajec (2011)
Amanita rubescens	0.58	4.88	1.64	0.61	10.3	Ribeiro et al. (2008a)
Boletus edulis	1.24	1.35	3.49	2.89	8.97	Tsai et al. (2008)
	0.75	10.2	2.47	0.84	22.7	Ribeiro et al. (2008a)
Calocybe gambosa	39.4	19.1	5.86	5.46	69.8	Beluhan and Ranogajec (2011)
	25.9	11.7	7.41	4.20	49.2	Beluhan and Ranogajec (2011)
Cantharellus cibarius	0.89	2.45	0.73	0.86	7.37	Ribeiro et al. (2008a)
	30.1	14.3	10.2	5.74	60.3	Beluhan and Ranogajec (2011)
Craterellus cornucopioides	45.9	7.08	7.85	4.71	65.5	Beluhan and Ranogajec (2011)
Flammulina velutipes	7.63	7.52	7.00	5.72	27.9	Beluhan and Ranogajec (2011)
Macrolepiota procera	33.8	8.19	6.68	4.13	52.8	Beluhan and Ranogajec (2011)
Pleurotus ostreatus	41.3	13.8	6.24	4.65	66.0	Beluhan and Ranogajec (2011)
Russula cyanoxantha	0.58	4.97	3.62	0.44	12.2	Ribeiro et al. (2008a)
Suillus granulatus	0.17	2.00	1.83	0.29	5.21	Ribeiro et al. (2008a)
Suillus luteus	1.58	4.66	1.43	0.39	8.94	Ribeiro et al. (2008a)
Terfezia claveryi	0.17	0.50	2.74	0.37	3.78	Kivrak (2015)
Terfezia olbiensis	0.16	0.43	2.37	0.41	3.37	Kivrak (2015)
Tricholoma equestre	1.06	10.3	1.72	3.16	20.3	Ribeiro et al. (2008a)
Cultivated species						
Agaricus bisporus	34.1	51.1	4.67	5.15	95.0	Kim et al. (2009)
	20.9	23.2	10.1	5.11	59.3	Liu et al. (2014)
	9.00	6.42	23.3	5.48	44.2	Pei et al. (2014)
Agaricus subrufescens	4.40	2.62	4.73	3.16	14.9	Tsai et al. (2008)
	28.7	49.4	10.7	4.90	93.7	Kim et al. (2009)
	2.97	7.00	7.85	4.90	22.7	Li et al. (2014b)

Agrocybe cylindracea	3.12	1.13	3.25	2.04	9.54	Tsai et al. (2008)
	3.62	3.86	5.01	0.72	13.2	Li et al. (2014b)
Clitocybe maxima caps	8.89	8.50	17.0	3.26	37.6	Tsai et al. (2009)
stipes	5.20	4.22	6.71	1.47	17.6	Tsai et al. (2009)
Coprinus comatus	4.99	1.70	2.71	0.44	5.93	Li et al. (2014b)
Flammulina velutipes white	1.57	10.5	6.38	0.76	19.2	Yang et al. (2001)
yellow	7.06	13.6	9.78	1.03	31.5	Yang et al. (2001)
	34.4	46.2	6.58	7.20	94.4	Kim et al. (2009)
Hypsizygus marmoreus white	16.1	23.8	37.3	18.7	95.9	Lee et al. (2009)
	11.2	7.63	13.2	4.58	36.6	Wu et al. (2015)
Lentinula edodes	1.71	7.77	2.53	0.51	12.5	Yang et al. (2001)
	17.7	9.83	1.65	2.37	31.6	Kim et al. (2009)
caps	2.34	11.7	4.77	2.10	23.6	Chen et al. (2015)
stipes	1.58	6.30	2.02	0.83	12.1	Chen et al. (2015)
	5.26	2.24	3.71	1.31	12.5	Li et al. (2014c)
Pleurotus cystidiosus	1.21	5.01	0.74	0.37	7.33	Yang et al. (2001)
	0.97	1.66	1.23	0.24	4.09	Li et al. (2014b)
Pleurotus eryngii	8.63	19.8	3.70	3.18	35.3	Kim et al. (2009)
	1.08	1.70	2.71	0.44	5.93	Li et al. (2014b)
	2.70	3.26	6.38	1.34	13.7	Li et al. (2015)
Pleurotus ferulae	1.76	2.69	4.24	1.79	10.5	Tsai et al. (2009)
Pleurotus ostreatus	0.84	2.25	0.78	0.21	4.08	Yang et al. (2001)
	43.3	42.1	8.56	3.96	97.9	Kim et al. (2009)
Sparassis crispa	2.14	3.53	5.23	1.62	12.5	Tsai et al. (2009)
	16.8	13.4	2.20	1.61	34.0	Kim et al. (2009)

MSG-like (monosodium glutamate-like): aspartic acid + glutamic acid.
Sweet: alanine + glycine + serine + threonine + proline.
Bitter: arginine + histidine + valine + leucine + isoleucine + methionine + phenylalanine.
Tasteless: lysine + tyrosine + cysteine.
Note: Although values of total free amino acids cited by Ribeiro et al. (2008a) also comprise asparagine, glutamine, tryptophan, and threonine, data of other authors comprise only the free amino acids listed here.

Table 3.2 Content of total and flavor 5′-nucleotides (mg g^{-1} dry matter) affecting taste of fresh mushrooms and equivalent umami concentration (EUC; g monosodium glutamate 100 g^{-1} dry matter)

Species	Total 5′-nucleotides	Flavor 5′-nucleotides	EUC	Reference
Wild-growing species				
Agaricus campestris	3.91	1.29	992	Beluhan and Ranogajec (2011)
Boletus edulis	2.76	2.01	10.5	Tsai et al. (2008)
	4.65	1.63	1186	Beluhan and Ranogajec (2011)
Calocybe gambosa	7.07	1.01	641	Beluhan and Ranogajec (2011)
Cantharellus cibarius	2.01	0.38	249	Beluhan and Ranogajec (2011)
Craterellus cornucopioides	35.4	13.9	121	Beluhan and Ranogajec (2011)
Flammulina velutipes	7.96	1.05	73.8	Beluhan and Ranogajec (2011)
Macrolepiota procera	4.55	0.52	318	Beluhan and Ranogajec (2011)
Morchella elata	20.4	5.32	79.5	Beluhan and Ranogajec (2011)
Pleurotus ostreatus	6.04	3.43	151	Beluhan and Ranogajec (2011)
Cultivated species				
Agaricus bisporus	12.9	7.02	–	Liu et al. (2014)
	5.35	0.68	751	Pei et al. (2014)
Agaricus subrufescens	8.00	5.15	136	Tsai et al. (2008)
	3.58	1.47	88.4	Li et al. (2014b)
Agrocybe cylindracea	8.56	2.44	46.7	Tsai et al. (2008)
	2.32	0.62	45.4	Li et al. (2014b)

Clitocybe maxima caps	1.89	0.54	–	Tsai et al. (2009)
stipes	7.59	3.54	–	Tsai et al. (2009)
Coprinus comatus	3.79	0.75	86.9	Li et al. (2014b)
Flammulina velutipes white	13.0	8.60	–	Yang et al. (2001)
yellow	13.2	6.32	–	Yang et al. (2001)
Hypsizygus marmoreus	6.43	1.67	351	Lee et al. (2009)
white	5.21	2.36	282	Wu et al. (2015)
Lentinula edodes	24.2	11.6	–	Yang et al. (2001)
caps	8.11	2.16	–	Chen et al. (2015)
stipes	3.78	0.86	–	Chen et al. (2015)
Pleurotus cystidiosus	13.9	5.52	–	Yang et al. (2001)
	1.87	0.67	13.3	Li et al. (2014b)
Pleurotus eryngii	1.68	0.78	11.2	Li et al. (2014b)
	26.8	0.50	30.0	Li et al. (2015)
Pleurotus ferulae	6.79	3.05	–	Tsai et al. (2009)
Pleurotus ostreatus	15.8	6.09	–	Yang et al. (2001)
	5.64	2.48	–	Tsai et al. (2009)

Flavor 5′-nucleotides: 5′-guanosine monophosphate + 5′-inosine monophosphate + 5′-xanthosine monophosphate.

carboxylic acids. Typical mushroom flavor (smell, aroma) consists of both volatile and nonvolatile components. Flavor is an important factor in the popularity of mushroom consumption.

3.1.1 Taste

Numerous mushroom compounds participate in the four basic tastes: sour, sweet, bitter, and salty. In particular, carboxylic acids (Section 3.3, Table 3.3) contribute to sour taste, and sugars, polyols (Section 2.4.1, Table 2.6), and several free amino acids (alanine, glycine, serine, and threonine) form a sweet taste. It is difficult to evaluate and generalize data regarding amino acids from Table 3.1. The available contents determined by different laboratories vary up to one order of magnitude in some species (eg, in *Boletus edulis*, *Agaricus subrufescens*, or *Pleurotus ostreatus*). A greater impact on sweet taste is caused by disaccharide trehalose and polyol mannitol occurring at considerably higher contents than the free amino acids. The most numerous group of free amino acids with a bitter taste comprises arginine, histidine, isoleucine, leucine, methionine, phenylalanine, and valine. The differences in their content reported by different laboratories are not as wide as those in the previous group. The highest content was observed in cultivated *Hypsizygus marmoreus* and among wild species in *Cantharellus cibarius*. Sweet compounds can probably mask the bitterness of the bitter components.

The fifth taste perceived in the mouth is the umami taste. It is also called the palatable taste or the perception of satisfaction, which is induced or enhanced by monosodium glutamate (MSG) and three 5′-nucleotides, namely monophosphates of guanosine, inosine, and xanthosine. 5′-Guanosine monophosphate elicits a meaty flavor that is much stronger than MSG. Two free amino acids, glutamic acid and aspartic acid, cause a palatable MSG-like taste of mushrooms. However, glutamine and asparagine produce sour and sweet tastes, respectively, not the umami taste (Kawai et al., 2012). According to the data collated in Table 3.1, great differences occur in the content of MSG-like tasting amino acids

Table 3.3 Content of aliphatic acids (mg 100g^{-1} dry matter) in selected mushroom species

Species	Oxalic acid	Succinic acid	Fumaric acid	Malic acid	Citric acid	Quinic acid	Reference
Wild-growing							
Agaricus albertii	620	–	160	1390	710	650	Reis et al. (2014a)
Agaricus bitorquis	4050	–	230	4400	ND	ND	Glamočlija et al. (2015)
Agaricus campestris	3470	–	650	4440	2390	ND	Glamočlija et al. (2015)
Agaricus macrosporus	260	–	200	1740	360	2590	Glamočlija et al. (2015)
Agaricus urinascens var. *excellens*	870	–	100	760	500	–	Reis et al. (2014a)
Amanita cesarea	170	–	320	1100	–	–	Fernandes et al. (2015)
Amanita curtipes	270	–	260	1500	–	–	Fernandes et al. (2015)
Amanita rubescens	149	12.8	206	1180	1360	–	Ribeiro et al. (2008b)
Armillariella mellea	–	–	–	1685	1790	–	Ayaz et al. (2011)
Boletus appendiculatus	2080	–	30	8570	ND	–	Barros et al. (2013)
Boletus edulis	66	6.4	3.1	882	217	–	Ribeiro et al. (2008b)
	–	–	–	493	2150	–	Ayaz et al. (2011)
	602	–	221	1730	ND	–	Barros et al. (2013)
	650	–	22	ND	4100	ND	Fernandes et al. (2014)
Boletus regius	170	–	70	ND	3320	–	Leal et al. (2013)
Calocybe gambosa	1190	–	51	2440	ND	–	Barros et al. (2013)
Cantharellus cibarius	131	–	163	3870	1200	–	Barros et al. (2013)
Coprinus comatus	492	–	848	2030	ND	–	Barros et al. (2013)
	680	–	650	4080	11,840	3370	Stojković et al. (2013)
	–	1540	5400	2390	6000	–	Li et al. (2014b)

(*Continued*)

Table 3.3 Content of aliphatic acids (mg 100 g^{-1} dry matter) in selected mushroom species (Continued)

Species	Oxalic acid	Succinic acid	Fumaric acid	Malic acid	Citric acid	Quinic acid	Reference
Craterellus cornucopioides	329	–	259	2780	ND	–	Barros et al. (2013)
Gyromitra esculenta	130	–	360	690	1460	–	Leal et al. (2013)
Hydnum repandum	–	–	–	309	670	–	Ayaz et al. (2011)
Lactarius deliciosus	511	–	114	2330	ND	–	Barros et al. (2013)
Laetiporus sulphureus	2660	–	250	–	1240	160	Petrović et al. (2014)
Lepista nuda	4340	–	68	869	ND	–	Barros et al. (2013)
Macrolepiota procera	–	–	–	1940	4090	–	Ayaz et al. (2011)
	1330	–	41	969	2640	–	Barros et al. (2013)
	440	–	200	2800	400	1000	Fernandes et al. (2014)
Morchella esculenta (Portugal)	32.3	–	47.8	199	ND	ND	Heleno et al. (2013)
(Serbia)	32.7	–	17.4	ND	233	43.6	Heleno et al. (2013)
Pleurotus eryngii	250	–	550	5150	430	ND	Reis et al. (2014a)
Russula cyanoxantha	–	–	727	5000	361	–	Ribeiro et al. (2008b)
Russula delica	210	–	114	2280	870	1800	Fernandes et al. (2014)
Russula virescens	780	–	230	2710	550	–	Leal et al. (2013)
Suillus granulatus	626	40.6	793	716	1320	–	Ribeiro et al. (2008b)
from Portugal	3350	–	920	ND	ND	360	Reis et al. (2014b)
from Serbia	420	–	1310	940	1770	ND	Reis et al. (2014b)

Species							References
Suillus variegatus	2460	–	22	383	ND	–	Barros et al. (2013)
Termitomyces robustus	690	–	10	–	–	–	Obodai et al. (2014)
Tricholoma portentosum	426	–	502	6490	1900	–	Barros et al. (2013)
Cultivated							
Agaricus bisporus	–	3810	53	2570	3080	–	Pei et al. (2014)
	580	–	Traces	1300	–	–	Stojković et al. (2014)
	3730	–	280	3820	ND	ND	Glamočlija et al. (2015)
Agaricus subrufescens	–	2670	2870	1600	11,310	–	Li et al. (2014b)
Auricularia auricula-judae	–	270	–	10	–	–	Obodai et al. (2014)
Hypsizygus marmoreus white	–	8790	51	–	704	–	Wu et al. (2015)
Lentinula edodes, caps	–	17,410	319	2660	5910	–	Chen et al. (2015)
stipes	–	6840	154	771	2760	–	Chen et al. (2015)
Lentinula squarrosulus	–	295	–	110	–	–	Obodai et al. (2014)
Pleurotus eryngii	–	2930	156	529	4360	–	Li et al. (2014b)
	–	–	220	5218	ND	–	Li et al. (2015)
Pleurotus ostreatus	–	290	–	180	–	–	Obodai et al. (2014)
Pleurotus sajor-caju	–	320	–	180	–	–	Obodai et al. (2014)
Pleurotus tuber-regium	–	690	–	10	–	–	Obodai et al. (2014)

ND, content below detection limit.

in the values reported by different laboratories for several species, namely *B. edulis*, *A. subrufescens*, *Lentinula edodes*, and *P. ostreatus*. It is thus difficult to deduce species with typically high or low levels of MSG-like taste.

Due to the synergistic effects of MSG-like components with flavor 5′-nucleotides, which may greatly increase the umami taste of mushrooms, Yamaguchi et al. (1971) established the equivalent umami concentration (EUC) with an equation and constants for its calculation. The EUC, expressed as grams of MSG $100 \, g^{-1}$ dry matter (DM), is the concentration of the MSG equivalent to the umami intensity given by a mixture of MSG and 5′-nucleotides. Mau (2005) grouped mushroom EUC values into four levels: >1000, 100–1000, 10–100, and <10 g $100 \, g^{-1}$ DM. The respective values for MSG are >10, 1–10, 0.1–1, and <0.1 g g^{-1} DM. As shown in Table 3.2, wild-growing mushrooms seem to have considerably higher EUC values than the cultivated species. This would confirm the higher palatability of commonly known wild species (eg, in the Central Europe). Nevertheless, the extreme difference in the EUC values in *B. edulis* demands further data from more laboratories.

A comprehensive review of the umami ingredients in edible mushrooms with numerous cited references has published Zhang et al. (2013). The umami taste of edible mushrooms was widely investigated in recent years and has shown a promising potential for use in the food spices industry. The presence of umami ingredients and their quantities in mushrooms are influenced by many factors including species, maturity stage, part of the mushroom (cap or stipe), storage time, and preservation treatment. The content of umami ingredients in edible mushrooms changes greatly with maturity stage and mushroom species. It is not yet possible from the available results to conclude and recommend the right maturity stage for harvest to get mushrooms with maximum umami ingredients. Thus, a preharvest analysis would be vital to decide the harvest time. However, the literature data indicate that maximum EUC level can be at low yield in some species.

From the fragmentary data, it seems that the caps are higher in the EUC than are the stipes. Moreover, each mushroom species has various substrains that differ from each other in the level of umami taste compounds. The effect of strains was found to have a lower effect on the MSG-like amino acid content than on the umami 5′-nucleotides.

The content of MSG-like free amino acids decreased significantly during storage of *L. edodes* packaged under various modified atmospheres. This indicates that none of the tested packaging methods could maintain the typical taste of just harvested shiitake (Li et al., 2014c). Freezing was the most saving preservation method of MSG-like amino acids in blanched *Agaricus bisporus* as compared with the canned or salted variants (Liu et al., 2014). Low EUC levels in canned *A. bisporus*, *Flammulina velutipes*, and *Volvariella volvacea* were also reported by Chiang et al. (2006).

Both freeze drying and energy-saving combinations of freeze drying with microwave vacuum drying effectively preserved MSG-like components in sliced *A. bisporus* as compared with the fresh mushroom (Pei et al., 2014). 5′-Cytosine monophosphate highly prevailed among 5′-nucleotides in *Pleurotus eryngii*. Freeze-drying was the drying procedure most saving of 5′-nucleotides, followed by vacuum drying, whereas ambient drying, hot-air drying, and microwave drying caused higher losses. On the contrary, hot-air drying greatly elevated the content of glutamic acid and preserved aspartic acid better than the other drying methods (Li et al., 2015). The effects of four drying methods on nonvolatile taste components in white *H. marmoreus* were reported by Wu et al. (2015). Freeze drying was the most saving method, followed by hot-air drying, whereas both microwave vacuum drying and vacuum drying caused considerably higher losses. The content of some taste 5′-nucleotides in *A. bisporus* was lowered by gamma irradiation (Sommer et al., 2010). However, Tsai et al. (2014) reported an increase in the same species, particularly following a dose of 2.5 kGy. Heating generally causes a loss of both MSG-like amino acids and taste 5′-nucleotides. The losses decrease in the following order: autoclaving > boiling > microwave cooking (Zhang et al., 2013).

A role in the taste of mushrooms was also observed in some other minor constituents, such as gamma–aminobutyric acid (GABA) (see Section 4.7), as a chemical inducer of the mouth-drying and mouth-coating oral sensation (Rotzoli et al., 2006). Free amino acid L-theanine (*N*-ethyl-L-glutamine) with a favorable taste, typically occurring in tea leaves (*Camellia sinensis*), was observed in *Xerocomus badius.*

3.1.2 Flavor

Unique and varied flavors exhibited by numerous species are among the most valued mushroom characteristics. Pleasant smells include anise-like, almond-like, floral, or fruity aromas, whereas coal tar smell is among disagreeable odors. Typical smells help in the sensory identification of species together with morphologic features.

Various volatiles emitted from fruit bodies play a role as attractants of insects, which are likely to disseminate spores and help in the reproduction process. However, other authors suggest that some volatiles could attract predators of fungal insect pests as a defense mechanism. Highly diverse ecological functions of volatile sesquiterpenes reviewed thoroughly Kramer and Abraham (2012).

Numerous original works regarding mushroom flavor have been reviewed (Combet et al., 2006; Gross and Asther, 1989; Maga, 1981). The volatiles can be grouped according to their chemical structure as alcohols, aldehydes, ketones, sesquiterpene-like compounds, terpenes, and various further compounds (eg, acids, esters, and sulfur-containing and heterocyclic components). Tens of volatiles have been detected in a species, whereas the total number of identified compounds from various species is in the order of hundreds. Nevertheless, it is virtually impossible to identify volatiles in an intact fruit body. The reported data provide information on the "postharvest" situation. Moreover, volatile content and composition can change considerably during aging of the fruit bodies. Within the tested parameters in *A. bisporus*, the fruit body developmental stage, postharvest storage, tissue type, and tissue disruption had a major impact on the profile of volatiles, both qualitatively and quantitatively (Combet et al., 2009).

Eight-carbon volatile compounds are a key contributor to mushroom flavor. The main eight-carbon volatiles present in mushrooms are 1–octen–3–ol, 1–octen–3–one, 3–octanol, 3–octanone, and 1–octanol, with chemical structures provided in Fig. 3.1. Aliphatic 1–octen–3–ol, often called "mushroom alcohol," is the principal volatile contributing to the unique fungal flavor. It has two isomers (enantiomers), the R-(−) form (with a stronger flavor) and the S-(+) form. Zawirska-Wojtasiak (2004) determined 1.53–5.23 mg of 1–octen–3–ol $100 \, g^{-1}$ of fresh matter (FM) in several cultivated

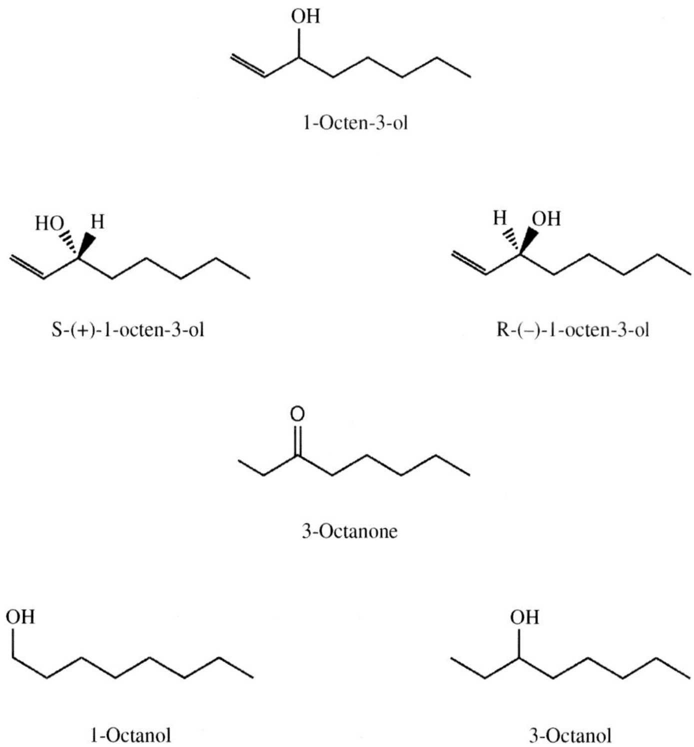

OH

1-Octen-3-ol

HO H

S-(+)-1-octen-3-ol

H OH

R-(−)-1-octen-3-ol

O

3-Octanone

OH

1-Octanol

OH

3-Octanol

Figure 3.1 Chemical structure of the main eight-carbon volatiles.

varieties of *A. bisporus*, with the proportion of R-(−)-enantiomer being at least 98.9%. The respective values for cultivated *Hericium erinaceum* were 1.18–3.46 mg $100 g^{-1}$ and more than 96.7%, for *P. ostreatus* were 6.62 mg $100 g^{-1}$ and 97.2%; however, values were only 1.21 mg $100 g^{-1}$ (95.5%) and 0.11 mg $100 g^{-1}$ (92.5%) for *L. edodes* and *Pholiota nameko*, respectively. Among wild-growing species, the highest level, 15.6 mg $100 g^{-1}$ (96.7%), was observed in *B. edulis*, while considerably lower both content and R(−) isomer proportion, 2.34 mg $100 g^{-1}$ and 82.7% were found in *X. badius*, and even lower contents were found in *Macrolepiota procera* (0.57 mg $100 g^{-1}$ and 98.3%). The results are in accordance with the highly valued strong aroma of cepes ("true boletes").

Several hypotheses have been formulated for the biochemistry of 1-octen-3-ol formation, but without having been fully proven. The eight-carbon volatiles evidently originate from linoleic acid (C_{18}; *cis, cis*-octadeca-9,12-dienoic acid); they are first oxidized and then cleaved. Linoleic acid belongs to minor fatty acids present in mushroom lipids (see Table 2.5); however, even such low content is sufficient for the formation of eight-carbon volatiles and ten-carbon compounds. At least two enzymes, lipoxygenase and hydroperoxide lyase, participate in the processes (for more information see the review by Combet et al., 2006).

However, the mentioned processes of linoleic acid oxidation and cleavage are more than a mushroom crop quality issue. Hydroperoxides (−O−OH) of the 18:2 polyunsaturated fatty acids and the produced eight-carbon volatiles are named oxylipins. In most fungi, linoleic acid is oxidized to form a 10-hydroperoxide intermediate, which is then cleaved to an eight-carbon volatile and ten-carbon oxoacid. 10-Oxodecanoic acid possesses hormone-like properties toward growth of the mushroom stipe and development of fungal structures. It is suggested that both 1-octen-3-ol and 10-oxodecanoic acid could work together to regulate the transition between vegetative and reproductive growth. Fungal oxylipins are thus at the crossroads of several biologically significant domains (Combet et al., 2006). Within such functions, volatile oxylipins act as attractants or repellents for specialist fungivores (Holighaus et al., 2014).

Surprisingly, 1-octen-3-ol was reported to exert toxicity in humans via disruption of dopamine homeostasis. It may thus represent a naturally occurring agent involved in Parkinson's disease (Inamdar et al., 2013).

Among volatiles of other chemical structures, *p*-anisaldehyde is characteristic for *Clitocybe odora*, and benzyl alcohol and benzaldehyde produce an almond-like aroma such as in dried *A. subrufescens*. The valued flavor of dried *L. edodes* is caused by lenthionine (1,2,3,6,6-pentathiepane), which is produced together with numerous other minor sulfur-containing compounds from lentinic acid, a dipeptide (Fig. 3.2), by sequence of enzyme-catalyzed reactions. The smell is absent in fresh shiitake mushrooms and weak after drying, and the amount of lenthionine in the dried fruit bodies increases during the rehydration and cooking processing (Hiraide et al., 2010).

De Pinho et al. (2008) tested the correlation between the volatiles and overall aroma of 11 wild edible species and divided them into three groups. *Tricholoma equestre, Amanita rubescens, C. cibarius,* and *Suillus bellinii* are rich in eight-carbon volatiles; *Fistulina hepatica, Suillus granulatus, Suillus luteus,* and *Russula*

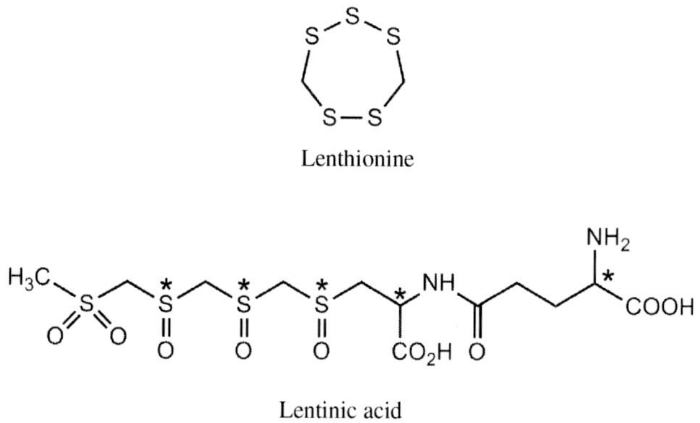

Figure 3.2 Chemical structure of lenthionine, which causes the valued flavor of dried *Lentinula edodes* (shiitake) and its precursor, lentinic acid.

cyanoxantha are rich in terpenic volatile compounds; and *B. edulis*, *Hygrophorus agathosmus*, and *Tricholomopsis rutilans* are rich in methional (3-methylthiopropanal). Methional was observed as the main aroma compound in *Pleurotus salmoneostramineus* (Usami et al., 2014).

According to Wang and Marcone (2011), aroma profiles of truffles vary among species. Eight-carbon volatiles are not typical for the group. Many widely variable chemical compounds were detected from alcohols and aldehydes with various lengths of carbon chains to heterocyclic and sulfur-containing compounds.

The flavor of air-dried and cooked mushrooms is much more complex and stronger than that of raw mushrooms. In addition to the eight-carbon volatile formation, numerous products of the Maillard reaction are produced (eg, heterocyclic compounds). As indicated by Grosshauser and Schieberle (2013), the characteristic odorants of raw mushrooms, 1-octen-3-one and 1-octen-3-ol, do not contribute much to the aroma of mushrooms after thermal processing. Because enzymes present in the raw cut mushroom tissue are able to generate odorants 3-methylbutanal and phenylacetaldehyde, which have a malty, honey-like smell, the overall aroma present after mushroom processing can clearly be influenced by the time lag during which the raw ground tissue is allowed to react.

Aprea et al. (2015) reported interesting information on changes in the composition of volatiles during storage of dried *B. edulis*. The fruit bodies were sliced and dried in a tunnel dryer and packaged commercially in polypropylene bags at residual moisture of approximately 12%. In total, 66 volatile compounds were identified. Alcohols, aldehydes, ketones, and monoterpenes diminished but carboxylic acids, pyrazines, and lactones increased during their 1-year shelf life. The work suggests that even short periods at temperatures higher than 25°C should be avoided to minimize cardinal changes in the volatiles profile.

Malheira et al. (2013) revealed 11 volatile compounds with evidential discriminating power for taxonomic and authentication purposes in 6 tested wild species. Such a concept was proven by a study by Zhou et al. (2015). Using electronic nose analysis, 88 volatiles in 8 dried edible species by gas chromatography-mass spectrometry and statistical methods were determined and the individual species were successfully distinguished. The composition of volatiles in commercial mushrooms could benefit a finger spectrum by using electronic nose analysis to identify mushroom species.

3.2 PIGMENTS

Mushroom pigmentation ranks among the main features used for species identification. The colors may vary during aging of a fruit body, and some of them undergo distinctive changes following tissue bruising. Moreover, the pigments may protect mushroom organisms from UV damage and bacterial attack or may play a role as insect attractants.

This topic has been very wide and demands an extensive knowledge of organic chemistry and biochemistry. Two recent comprehensive reviews are available. Zhou and Liu (2010) have focused primarily on chemical aspects of numerous groups of pigments, whereas Velíšek and Cejpek (2011) have dealt with the pigments typical for the individual mycological orders. Only basic information is given in this text.

Mushrooms contain pigments other than those dominating in higher plants. Chlorophylls and anthocyanins are lacking. Betalains, carotenoids, and other terpenoids occur only in some mushroom species. Quinones or similar conjugated structures form the great proportion of mushroom pigments. They have usually been classified according to four pathways of their biosynthesis, namely shikimate (chorismate), acetatemalonate (polyketide), mevalonate (terpenoid), and nitrogen-containing structures.

Numerous pigments are formed from their colorless precursors following bruising or mechanical damage of fruit bodies (eg, sliced or bruised). Under such conditions, a precursor and relevant catalyzing enzyme, so far separated within the tissue, come into contact together with the action of air oxygen.

Such changes cause great economic losses, particularly due to browning of white button mushrooms, *A. bisporus*, and the relevant research has thus been extensive. Both bruising-tolerant and bruising-sensitive strains exist. Tens of various phenolics were identified in the species. Some of them are oxidized under the catalysis of polyphenol oxidases to their respective quinones. The quinones undergo oxidative polymerization leading to high-molecular dark brown or black pigments, melanins. Four pathways for melanin synthesis have been proposed. Great attention has been given to the pathway, including hydroxylation of amino acid tyrosine to 3,4-dihydroxyphenylalanine (L-DOPA) catalyzed by the enzyme tyrosinase (for reviews see Jolivet et al., 1998; Ramsden and Riley, 2014). The following two phenolics were identified as compounds correlated with bruising sensitivity: γ-L-glutaminyl-4-hydroxybenzene and γ-L-glutaminyl-3, 4-dihydroxybenzene (Weijn et al., 2013).

3.3 ALIPHATIC ACIDS

Several aliphatic carboxylic acids have been reported in mushrooms. Data for selected wild-growing species are presented in Table 3.3, and information on numerous other species is available in the work by Barros et al. (2013). Malic, citric, and oxalic acids are most common, followed by fumaric acid (Fig. 3.3). Information on succinic acid and quinic acid has been very limited, and low levels of several other acids have been reported sporadically. The acids have various roles in mushroom physiology and participate in mushroom taste. As in other vegetables, the acids are probably also stable in mushrooms under various cooking conditions.

Data in Table 3.3 show significant differences in the reported contents of acids among and within species. The differences

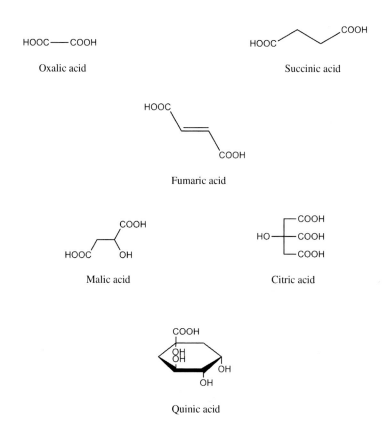

Figure 3.3 Chemical structure of the main aliphatic acids occurring in mushrooms.

appear to be up to one order of magnitude (eg, in *B. edulis*); therefore, it is difficult to draw a credible conclusion. Nonetheless, malic acid is usually the main aliphatic acid, with the content comprising up to several percent DM. The ambiguous data deal with citric acid. Although some works reported contents comparable with malic acid, other have reported nondetectable levels. Such differences are striking, such as in *Coprinus comatus* or *B. edulis*. Exogenous citric acid is known to extend mushroom shelf life and to prevent browning (see Section 3.2). It is an effective antioxidant that chelates transition metals (copper, iron) into nonfunctioning chemical forms.

Oxalic acid is present in foods as free and/or as salts called oxalates. Free acid, potassium, and sodium oxalates are water-soluble, but calcium oxalate is insoluble. Soluble dietary oxalates combine in the human body with calcium to crystals of calcium oxalate, which can irritate the gut and kidneys. Healthy adult individuals can safely consume in moderate level foods containing increased levels of oxalates (eg, spinach or rhubarb, up to approximately 250 mg day^{-1}). However, those with kidney disorders, gout, osteoporosis, or rheumatoid arthritis are advised to limit oxalate intake to less than 40 mg day^{-1}.

Soluble oxalates in wild species (Savage et al., 2002) have been reported at 29–40 mg 100 g^{-1} DM. Sembratowicz and Rusinek-Prystupa (2012) determined levels of 36–104 mg 100 g^{-1} DM in four related species from the Boletaceae family, with the lowest level in *S. luteus*. Caps had 1.6- to 3.1-times higher oxalate content than stipes. However, data in Table 3.3, originating from a Portuguese laboratory, provide considerably higher contents in most species, with some exceeding 1000 mg 100 g^{-1} DM (ie, more than approximately 100 mg 100 g^{-1} FM). Consumption of such species should thus be very limited in individuals who should not consume oxalates.

3.4 PHENOLIC COMPOUNDS

Phenolic compounds or phenolics comprise a large group of mushroom constituents with variable chemical structures, particularly phenolic acids. Total phenolic content (TPC) is commonly determined by the Folin-Ciocalteu's photometric assay and expressed as gallic acid equivalent (GAE). The usual TPC is between 0.1 and 0.6 g GAE 100 g^{-1} DM, and is only rarely more than 1.0 g 100 g^{-1} DM (eg, Beara et al., 2014; Dubost et al., 2007; Guo et al., 2012; Kolayli et al., 2012; Radzki et al., 2014; Woldegiorgis et al., 2014; Yildiz et al., 2015). For instance, Witkowska et al. (2011) determined TPC to be between 0.37 and 1.68 g GAE 100 g^{-1} DM in 16 wild species. Nowacka et al.

(2014) determined TPC to be between 0.01 and 0.32 g GAE 100 g^{-1} DM in 19 species commonly collected and consumed in the Central Europe. The highest level was observed in the genus *Boletus.*

Determination of TPC by the Folin–Ciocalteu assay can lead to overestimated values because the agent also reacts with other nonphenolic reducing compounds such are ascorbic acid, some sugars, and amino acids. Extraction of dried mushrooms with water yielded somewhat higher TPC levels than did extraction with ethanol (Radzki et al., 2014). Phenolics greatly participate in the antioxidant capacity of mushrooms (see Section 4.1).

Great differences in TPC occurred among years of harvest in cultivated *A. bisporus* and seven strains of *A. subrufescens.* The differences were higher than between the species harvested within 1 year (Geösel et al., 2011). A higher level of TPC was determined in *P. eryngii* harvested on day 10 than that harvested on days 12 or 15 after the initial fruit body formation (Lin et al. 2014). Within four tested drying methods of white *A. bisporus,* the losses of TPC were in the following order: sun drying > hot-air drying > freeze drying > microwave-vacuum drying (Ji et al., 2012). However, gamma irradiation up to 5 kGy in the same species had no significant effect on TPC (Sommer et al., 2009). Blanching of cultivated *P. ostreatus* fruit bodies in boiling water for 4 min decreased total polyphenol content from 0.49 to $0.12 \text{ g } 100 \text{ g}^{-1}$ DM. The loss was largely due to leaching into the blanching water (Lam and Okello, 2015).

Within the mushroom phenolics, the main interest has focused on phenolic acids. The data are presented in Table 3.4, and chemical structures of the main acids are given in Fig. 3.4. Phenolic acids are synthesized via the shikimate pathway from amino acids L-phenylalanine and L-tyrosine. The most documented are *p*-hydroxybenzoic, protocatechuic, *p*-coumaric, and gallic acids, with the usual contents being less than $5 \text{ mg } 100 \text{ g}^{-1}$ DM; levels more than $10 \text{ mg } 100 \text{ g}^{-1}$ DM have been sporadic. The contents of other acids presented in Table 3.4 seem to be very low.

Table 3.4 Content of phenolic acids (mg 100 g^{-1} dry matter) in selected mushroom species

Species	Hydroxybenzoic acids					Hydroxycinnamic acids				Reference
	p-Hydroxybenzoic	Proto-catechuic	Gallic	Vanillic	Syringic	o-Coumaric	p-Coumaric	Caffeic	Sinapic	
Wild-growing species										
Agaricus albertii	8.23	–	–	–	–	–	3.53	–	–	Reis et al. (2014a)
Agaricus arvensis	7.01	ND	–	ND	–	–	4.87	–	–	Barros et al. (2009)
Agaricus bisporus	2.56	ND	–	ND	–	–	ND	–	–	Barros et al. (2009)
Agaricus bitorquis	0.03	ND	ND	–	–	–	ND	–	–	Glamočlija et al. (2015)
Agaricus campestris	3.87	–	56.2	–	–	–	1.09	ND	–	Woldegiorgis et al. (2014)
Agaricus macrosporus	4.13	1.07	ND	–	–	–	0.68	–	–	Glamočlija et al. (2015)
	ND	ND	ND	–	–	–	ND	–	–	Glamočlija et al. (2015)
Agaricus silvicola	23.9	ND	–	ND	–	–	4.57	–	–	Barros et al. (2009)
Agaricus urinascens var. excellens	3.27	–	–	–	–	–	1.33	–	–	Reis et al. (2014a)

Species										Reference
Armillariella mellea	0.40	ND	–	–	–	–	ND	–	–	Vaz et al. (2011a)
	4.57	–	17.3	–	–	–	–	–	–	Guo et al. (2012)
	ND	0.22	–	ND	–	–	ND	–	0.38	Muszyńska et al. (2013c)
	ND	0.23	–	ND	–	–	ND	ND	–	Nowacka et al. (2014)
Boletus aestivalis	1.21	ND	–	–	–	–	ND	–	–	Heleno et al. (2011)
Boletus appendiculatus	1.34	ND	–	–	–	–	0.45	–	–	Heleno et al. (2011)
Boletus edulis	0.66	0.20	–	–	–	–	0.12	–	–	Heleno et al. (2011)
	0.19	0.75	–	ND	–	–	ND	–	ND	Muszyńska et al. (2013c)
Boletus regius	1.77	1.15	–	–	–	–	2.08	–	–	Leal et al. (2013)
Calocybe gambosa	3.84	0.26	–	–	–	–	0.40	–	–	Vaz et al. (2011a)
Cantharellus cibarius	0.23	0.15	–	0.33	–	–	ND	–	0.30	Muszyńska et al. (2013c)
Clitocybe geotropa	2.89	0.20	ND	ND	ND	0.02	0.52	ND	–	Kolayli et al. (2012)
Clitocybe odora	2.79	ND	–	–	–	–	0.18	–	–	Vaz et al. (2011a)

(Continued)

Table 3.4 Content of phenolic acids (mg 100 g^{-1} dry matter) in selected mushroom species (Continued)

Species	Hydroxybenzoic acids					Hydroxycinnamic acids				Reference
	p-Hydro-xybenzoic	Proto-catechuic	Gallic	Vanillic	Syringic	o-Coumaric	p-Coumaric	Caffeic	Sinapic	
Coprinus comatus	6.15	ND	–	–	–	–	0.18	–	–	Vaz et al. (2011a)
	0.09	–	ND	–	–	–	0.15	–	–	Stojković et al. (2013)
Craterellus cornucopioides	1.27	0.21	–	0.15	–	–	0.03	ND	ND	Nowacka et al. (2014)
Fistulina hepatica	4.19	6.76	–	–	–	–	ND	–	–	Vaz et al. (2011b)
Ganoderma lucidum	0.52	0.30	–	1.57	0.23	–	0.14	–	–	Yildiz et al. (2015)
Gyromitra esculenta	ND	3.74	–	–	–	–	ND	–	–	Leal et al. (2013)
Hericium erinaceum	0.24	–	Traces	0.61	0.02	–	0.60	–	–	Yildiz et al. (2015)
Hydnum repandum	ND	ND	–	–	–	–	ND	–	–	Vaz et al. (2011b)
Laccaria amethystea	1.77	ND	–	ND	–	–	ND	ND	–	Nowacka et al. (2014)
Laccaria laccata	ND	ND	–	ND	–	–	ND	ND	–	Nowacka et al. (2014)

Lactarius deliciosus	0.22	ND	–	0.002	0.005	0.27	0.03	0.003	0.004	Kalogeropoulos et al. (2013)
	ND	0.14	–	ND	–	–	ND	–	1.43	Muszyńska et al. (2013c)
Lactarius sanguifluus	0.21	0.003	–	0.003	0.006	0.24	0.03	0.03	0.004	Kalogeropoulos et al. (2013)
Lactarius semisanguifluus	0.17	ND	–	0.002	0.007	0.24	0.002	0.005	0.006	Kalogeropoulos et al. (2013)
Laetiporus sulphureus	0.05	6.74	–	–	–	–	ND	0.25	–	Woldegiorgis et al. (2014)
	0.08	ND	–	ND	–	–	0.02	ND	–	Nowacka et al. (2014)
Lecinum scabrum	0.05	ND	–	ND	–	–	0.05	ND	–	Nowacka et al. (2014)
Lentinula edodes	0.24	–	Traces	0.21	0.10	–	0.05	–	–	Yildiz et al. (2015)
Lycoperdon perlatum	0.37	ND	–	ND	–	–	0.19	ND	–	Nowacka et al. (2014)
Macrolepiota procera	ND	0.52	–	ND	–	–	ND	ND	–	Nowacka et al. (2014)
	0.11	0.08	–	–	–	–	0.19	–	–	Fernandes et al. (2014)
Marasmius oreades	0.16	ND	–	ND	–	–	ND	ND	–	Nowacka et al. (2014)

(Continued)

Table 3.4 Content of phenolic acids (mg 100 g^{-1} dry matter) in selected mushroom species (Continued)

Species	Hydroxybenzoic acids					Hydroxycinnamic acids				Reference
	p-Hydro-xybenzoic	Proto-catechuic	Gallic	Vanillic	Syringic	o-Coumaric	p-Coumaric	Caffeic	Sinapic	
Morchella esculenta from Portugal	0.10	0.24	–	–	–	–	0.01	–	–	Heleno et al. (2013)
from Serbia	0.10	0.06	–	–	–	–	ND	–	–	Heleno et al. (2013)
	0.35	1.72	0.08	–	–	–	Traces	–	–	Yildiz et al. (2015)
Pleurotus cornucopiae	ND	0.64	ND	ND	0.17	ND	ND	ND	–	Kolayli et al. (2012)
Pleurotus eryngii	3.81	–	–	–	–	–	ND	–	–	Reis et al. (2014a)
Pleurotus ostreatus	ND	0.26	ND	0.54	–	ND	0.01	ND	–	Kolayli et al. (2012)
Ramaria botrytis	1.40	34.3	–	ND	–	–	ND	–	–	Barros et al. (2009)
	5.00	10.8	19.2	–	–	–	–	–	–	Guo et al. (2012)
Rozites caperata	2.59	ND	–	ND	–	–	ND	ND	–	Nowacka et al. (2014)
Russula delica	0.02	ND	–	0.004	0.01	0.06	0.02	0.002	0.006	Kalogeropoulos et al. (2013)

Russula virescens	22.6	ND	—	—	—	—	ND	ND	—	Leal et al. (2013)
Sparassis crispa	0.10	ND	—	ND	—	—	ND	ND	—	Nowacka et al. (2014)
Suillus bellinii	0.13	0.05	—	0.004	0.004	0.29	0.02	0.004	0.01	Kalogeropoulos et al. (2013)
Suillus bovinus	5.90	15.2	17.2	—	—	—	—	—	—	Guo et al. (2012)
Suillus granulatus from Portugal	0.48	—	0.11	—	—	—	—	—	—	Reis et al. (2014b)
from Serbia	0.13	—	ND	—	—	—	—	—	—	Reis et al. (2014b)
Terfezia claveryi	1.70	1.55	—	1.15	0.21	—	0.76	ND	—	Kivrak (2015)
Terfezia olbiensis	1.81	2.16	—	0.98	0.21	—	1.40	ND	—	Kivrak (2015)
Termitomyces clypeatus	1.17	10.2	—	—	—	—	1.0	ND	—	Woldegiorgis et al. (2014)
Termitomyces robustus	2.43	—	—	—	—	—	0.24	—	—	Obodai et al. (2014)
Tremella fuciformis	32.3	—	—	—	—	—	3.0	—	—	Li. et al. (2014a)
Tuber aestivum	1.65	3.00	6.37	—	—	1.34	9.37	—	—	Villares et al. (2012)
Tuber indicum	ND	5.84	6.15	—	—	ND	ND	—	—	Villares et al. (2012)

(*Continued*)

Table 3.4 Content of phenolic acids (mg 100 g^{-1} dry matter) in selected mushroom species (Continued)

Species	Hydroxybenzoic acids					Hydroxycinnamic acids				Reference
	p-Hydroxybenzoic	Proto-catechuic	Gallic	Vanillic	Syringic	o-Coumaric	p-Coumaric	Caffeic	Sinapic	
Xerocomus badius	0.13	2.14	–	ND	–	–	1.39	–	0.15	Muszyńska et al. (2013c)
	ND	0.12	–	ND	–	–	ND	ND	–	Nowacka et al. (2014)
Cultivated species										
Agaricus bisporus white	0.05	<0.03	–	–	–	–	–	0.08	–	Mattila et al. (2001)
	ND	ND	6.28	–	–	–	0.23	–	–	Reis et al. (2012b)
brown	0.65	0.11	–	–	–	–	–	0.07	–	Mattila et al. (2001)
	ND	ND	ND	–	–	–	ND	–	–	Reis et al. (2012b)
	–	–	0.1	–	–	–	ND	–	–	Stojković et al. (2014)
	ND	ND	0.32	–	–	–	0.12	–	–	Glamočlija et al. (2015)
Agaricus subrufescens	–	–	ND	–	–	–	0.28	–	–	Stojković et al. (2014)
Auricularia auricola-judae	1.09	–	–	–	–	–	ND	–	–	Obodai et al. (2014)

Species										Reference
Coprinus comatus	—	—	ND	—	—	—	0.10	—	0.08	Stojković et al. (2013)
Lentinula edodes	—	<0.05	—	—	—	—	—	0.14	0.79	Mattila et al. (2001)
	—	—	ND	—	—	—	ND	0.04	0.16	Reis et al. (2012b)
	—	0.08	ND	—	—	—	6.12	—	0.16	Woldegiorgis et al. (2014)
Lentinula squarrosulum	—	—	0.10	—	—	—	—	—	1.48	Obodai et al. (2014)
Pholiota nameko	—	—	1.43	—	—	—	0.84	—	0.42	Gąsecka et al. (2015)
Pleurotus eryngii	—	—	0.10	—	—	—	ND	Traces	Traces	Reis et al. (2012b)
	—	—	1.60	—	—	—	—	—	0.28	Gąsecka et al. (2015)
Pleurotus ostreatus	—	—	0.08	—	—	ND	ND	0.08	0.16	Reis et al. (2012b)
	0.21	—	ND	—	—	—	—	0.25	0.36	Muszyńska et al. (2013c)
	—	—	0.21	—	—	—	—	—	1.56	Obodai et al. (2014)
	—	0.78	ND	—	—	—	—	1.30	0.13	Woldegiorgis et al. (2014)
	—	—	1.75	—	—	—	—	—	0.58	Gąsecka et al. (2015)

(Continued)

Table 3.4 Content of phenolic acids (mg 100 g^{-1} dry matter) in selected mushroom species (Continued)

Species	Hydroxybenzoic acids					Hydroxycinnamic acids				Reference
	p-Hydro-xybenzoic	Proto-catechuic	Gallic	Vanillic	Syringic	o-Coumaric	p-Coumaric	Caffeic	Sinapic	
Pleurotus sajor-caju	0.43	—	—	—	—	—	ND	—	—	Obodai et al. (2014)
Pleurotus tuber-regium	0.08	—	—	—	—	—	ND	—	—	Obodai et al. (2014)
Tuber aestivum[a]	0.07	ND	ND	—	—	—	—	—	—	Beara et al. (2014)
Tuber magnatum[a]	ND	0.02	ND	—	—	—	—	—	—	Beara et al. (2014)

ND, content below detection limit.
[a]Content in water extract obtained by maceration.

(A) Hydroxybenzoic acids

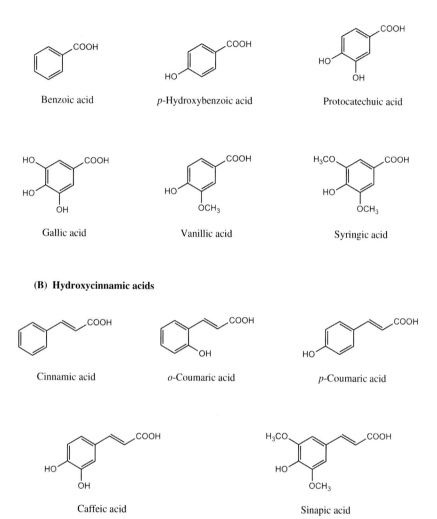

Figure 3.4 Chemical structure of the main phenolic acids occurring in mushrooms.

Information on further phenolic acids, such as gentisic, chlorogenic, ferulic, *p*-hydroxyphenylacetic, and 3,4-dihydroxyphenylacetic, has been scarce and only very low contents have been reported. Variegatic acid containing four phenolic groups in two pairs of *o*-positions was indicated as the main antioxidant compound in *B. edulis* and *Leccinum aurantiacum* (Vidović et al., 2010). The fate of plant and mushroom phenolic acids after their absorption in the gastrointestinal tract was extensively reviewed by Heleno et al. (2015). The acids circulate in human plasma in the form of conjugates, glucuronides, and methylated and sulfated derivatives. Such changes in the structures may decrease or increase their biological activities as compared to the initial compounds.

Data on mushroom phenols have also been sporadic. Tyrosol, vanillin, and resveratrol from the group of stilbenes were observed at negligible levels (Kalogeropoulos et al., 2013).

In general, it has been assumed that flavonoids can be biosynthesized only by plants. Barros et al. (2009) detected no flavonoids in 16 species of Portuguese wild mushrooms. However, several of approximately 5000 known polyphenolic flavonoids were reported in mushrooms, namely kaempferol, quercetin, morin, myricetin, chrysin, naringenin, and amentoflavone. Their individual contents were very low (usually less than $0.5\,mg\ 100\,g^{-1}$ DM) (Akyüz et al., 2012; Kalogeropoulos et al., 2013). The proportion of flavonoids within total phenolics seems to be low, as do their antioxidant and anti-inflammatory contributions.

3.5 STEROLS

Besides ergosterol and provitamin D_2 described in Section 2.6.1, several other sterols were observed in edible mushroom fruiting bodies. Kalogeropoulos et al. (2013) reported ergosta-5,7-dienol, ergosta-7,22-dienol, ergosta-7-enol, lanosterol, lanosta-8,24-dienol, and 4α-methylzymosterol in five species from Greece. The contents were lower by one to up to three orders of magnitude than those of ergosterol (see Table 2.13). No fewer than 28 new sterols have been identified by Yaoita et al. (2014) in various Japanese medicinal mushroom species. Similarly, 23 sterols were identified at levels of $190–320\,mg\ 100\,g^{-1}$ DM in fruit bodies of five truffle species of the genus *Tuber*. In addition to ergosterol, four phytosterols (ie, brassicasterol, campesterol, stigmasterol, and β-sitosterol) with beneficial health effects (Tang et al., 2012) were prevalent. Sterols commonly occur as esters bound with fatty acids.

Irradiation of mushrooms with UV-B light for vitamin D_2 formation initiates further products of ergosterol photodegradation, particularly lumisterol$_2$ and tachysterol$_2$. These compounds were detected in irradiated white *A. bisporus* (Kalaras et al., 2012) and *P. ostreatus* (Krings and Berger, 2014) at a level of several mg $100\,g^{-1}$ DM.

Because both dietary phytosterols and mycosterols are able to reduce cholesterol levels in blood serum, two extraction techniques were tested for sterol isolation from both the complete fruit bodies of *A. bisporus* and the lower part of the stipe, which is usually discarded after the harvest (Gil-Ramírez et al., 2013). Supercritical fluid extraction with carbon dioxide produced concentrates with 60% sterols, whereas pressurized liquid extraction with ethanol yielded extracts with 2.9–5% sterols.

Ergosterol peroxide, a derivative with cytotoxic and other auspicious biological activities, is found in mushrooms. It is believed to be either an intermediate in the hydrogen peroxide-dependent enzymatic oxidation in the biosynthetic pathway of steroidal dienes or a product of reactive oxygen species (free radicals) detoxification. The contents are not negligible; Krzyczkowski et al. (2009) reported 29.3, 17.3, and 12.6 mg $100\,g^{-1}$ DM in *B. edulis*, *Suillus bovinus*, and *X. badius*, respectively. Nevertheless, elevated contents are probably caused by ergosterol oxidation by atmospheric oxygen catalyzed with sunshine.

3.6 INDOLE COMPOUNDS

Nonhallucinogenic indole derivatives fulfill numerous activities in human physiology, such as neurotransmitters or their precursors, regulators of diurnal cycle, or anti-inflammatory compounds. Amino acid L-tryptophan is, biochemically, the key compound of the group. It is the precursor of the hormones serotonin and melatonin. Via biogenic amine tryptamine, a decarboxylation product of L-tryptophan, indoleacetic acid and its nitrile, which are

Compound	Position 3	Position 5
Tryptophan	$-CH_2-CH(NH_2)-COOH$	$-H$
5-Hydroxytryptophan	$-CH_2-CH(NH_2)-COOH$	$-OH$
5-Methyltryptophan	$-CH_2-CH(NH_2)-COOH$	$-CH_3$
Tryptamine	$-CH_2-CH_2-NH_2$	$-H$
Serotonin	$-CH_2-CH_2-NH_2$	$-OH$
Melatonine	$-CH_2-CH_2-NH-CO-CH_3$	$-OCH_3$
Indoleacetic acid	$-CH_2-COOH$	$-H$
Indoleacetonitrile	$-CH_2-C\equiv N$	H

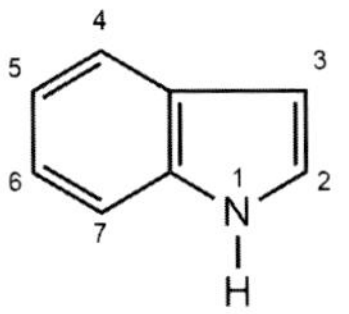

Figure 3.5 Indole-containing compounds detected in mushrooms.

known as plant growth regulators, are produced. Chemical structure is given in Fig. 3.5. Tryptophan is very sensitive to oxygen, particularly under acidic conditions. Among its oxidation products are kynurenine, which has a mutagenic effect.

Available data for fresh mushrooms are presented in Table 3.5. Moreover, low contents of indoleacetamide and 5-methyltryptophan were determined in several species. The contents of the amino acid L-tryptophan are significantly lower than those reported in Table 2.4. Nevertheless, the data of Table 3.5 represent unbound tryptophan, with the highest level in *S. bovinus.* The content of 5-hydroxytryptophan is more than 15 mg $100\,g^{-1}$ DM only in three analyzed species. A high level of serotonin in several wild-growing species is notable, and a high level of kynurenine-sulfate was observed in *Lactarius deliciosus.* The other compounds seem to be of minor importance.

Indole compounds are not stable under higher temperatures and easily break down. Tryptophan is the most stable within the group, and is resistant up to 90°C. The effects of boiling the dried powder of several species in water for 60 min on indole compound

Table 3.5 Content of indole compounds (mg 100 g^{-1} dry matter) in selected fresh mushroom species

Species	L-Tryptophan	5-Hydroxytryptophan	Tryptamine	Melatonin	Serotonin	Kynurenic acid	Kynurenine sulphate	Indoleacetic acid	Indoleacetonitrile	Reference
Wild-growing species										
Armillariella mellea	4.47	ND	2.74	ND	2.21	ND	ND	ND	ND	Muszyńska et al. (2011b)
Boletus edulis	0.39	0.18	1.17	0.68	10.1	ND	ND	0.04	ND	Muszyńska and Sułkowska-Ziaja (2012)
Cantharellus cibarius	0.01	0.02	0.01	0.14	29.6	ND	4.81	ND	0.01	Muszyńska et al. (2011a)
	0.02	0.01	0.02	0.11	17.6	ND	3.62	ND	0.02	Muszyńska et al. (2013b)
Lactarius deliciosus	ND	0.25	ND	1.29	18.4	ND	39.2	2.04	0.62	Muszyńska et al. (2011a)
Leccinum aurantiacum	ND	0.02	1.05	0.08	31.7	ND	ND	0.89	0.45	Muszyńska et al. (2011a)
Leccinum scabrum	9.56	ND	ND	ND	14.0	–	–	–	–	Muszyńska et al. (2013a)
Macrolepiota procera	3.47	22.9	0.92	0.07	ND	–	–	–	–	Muszyńska et al. (2013a)
Sarcodon imbricatus	13.0	–	22.1	2.13	52.0	–	–	–	–	Sułkowska-Ziaja et al. (2014)

(Continued)

Table 3.5 Content of indole compounds (mg $100\,g^{-1}$ dry matter) in selected fresh mushroom species (Continued)

Species	L-Trypto-phan	5-Hydro-xytrypto-phan	Trypt-amine	Mela-tonin	Serotonin	Kynurenic acid	Kynure-nine sulphate	Indole-acetic acid	Indole-acetonit-rile	Reference
Suillus bovinus	25.9	15.8	3.15	ND	ND	–	–	–	–	Muszyńska et al. (2013a)
Tricholoma flavovirens	2.85	0.59	2.01	ND	0.18	ND	ND	ND	ND	Muszyńska et al. (2009)
Xerocomus badius	0.68	ND	0.47	ND	0.52	1.57	1.96	ND	ND	Muszyńska et al. (2009)
Cultivated species										
Agaricus bisporus	0.39	ND	0.06	0.11	5.21	6.21	ND	0.19	ND	Muszyńska et al. (2011a)
Auricularia polytricha	0.16	7.32	2.77	ND	ND	–	–	–	–	Muszyńska et al. (2013a)
Lentinula edodes	0.58	24.8	0.04	0.13	1.03	–	–	–	–	Muszyńska et al. (2013a)
Pleurotus ostreatus	ND	2.08	0.91	ND	6.52	ND	ND	0.01	0.21	Muszyńska and Sułkowska-Ziaja (2012)

ND, content below detection limit.

content changes and composition are reported by Muszyńska and Sułkowska-Ziaja (2012) and Muszyńska et al. (2013a). The compound profiles changed considerably. The tryptophan level significantly increased, whereas no 5-hydroxytryptophan, serotonin, indoleacetic acid, and kynurenine derivatives were detected; the changes in melatonin content were equivocal. Moreover, products of thermal changes, indole and 5-methyltryptophan, were newly present.

From a nutritional point of view, knowledge of indole compounds released from mushroom fruit bodies in the gastrointestinal tract is needed. Muszyńska et al. (2015) tested the release rate in artificial stomach juice. The total released indole compounds were 262, 123, and 1.4 mg $100\,g^{-1}$ DM from cultivated *A. bisporus*, wild *X. badius*, and *C. cibarius*, respectively. 5-Hydroxy-L-tryptophan was the dominant compound.

Overall, mushrooms seem to be rich in indole compounds, particularly serotonin. Due to their observed instability during boiling, great changes in their profile can also be expected under other thermal treatments. Mushrooms seem to be a promising source of precursors of indole derivatives and possess various physiological activities in humans.

3.7 PURINE COMPOUNDS

Gout is a serious disease characterized by abnormally high levels of uric acid in the body, resulting in the formation and deposition of urate crystals in the joints and kidneys. Daily dietary intake of purines by patients with gout and hyperuremia should be less than 400 mg. From this point of view, total purine determined in seven cultivated mushroom species seems to be acceptable. Total contents were 142, 98.5, 49.4, 20.8, 20.8, 13.4, and 9.5 mg $100\,g^{-1}$ FM in *P. ostreatus*, *Grifola frondosa*, *F. velutipes*, *H. marmoreus*, *L. edodes*, *P. eryngii*, and *P. nameko*, respectively. Adenine and guanine were prevalent, whereas proportions of hypoxanthine and xanthine (chemical structure shown in Fig. 3.6) were limited (Kaneko et al.,

Compound	Position 2	Position 4
Adenine	$-H$	$-NH_2$
Guanine	$-NH_2$	$-OH$
Hypoxanthine	$-H$	$-OH$
Xanthine	$-OH$	$-OH$

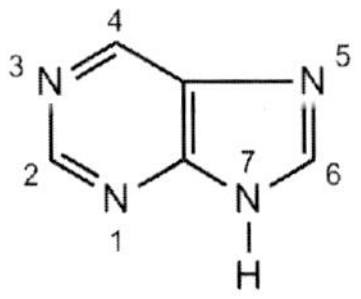

Figure 3.6 Purine compounds contributing to mushroom taste.

2008). Yuan et al. (2008) reported slightly lower adenine in several Chinese mushrooms used in folk medicine. They reported that the highest levels of total nucleosides and nucleobases were observed in the spore-forming part of fruit bodies.

In fresh mushrooms, purines are present in high-molecular nucleic acids. During the storage period the content of nucleic acids considerably decreased in favor of low-molecular nucleotides, nucleosides, and free bases. During cooking, the low-molecular compounds were leached into the water. However, cooking preserved a higher proportion of purine compounds than drying (Lou and Montag, 1994).

In addition, an inhibitor of xanthine oxidase isolated from *P. ostreatus* showed antigout activity in experimental rats. Xanthine oxidase is a rate-limiting enzyme in the biosynthesis of uric acid and catalyzes the oxidation of hypoxanthine and xanthine to uric acid. The inhibitor was identified as a tripeptide with the following sequence of amino acids: phenylalanine–cysteine–histidine (Jang et al., 2014).

3.8 BIOGENIC AMINES AND POLYAMINES

Biogenic amines (Fig. 3.7) are mainly produced in food materials by bacterial decarboxylation of free amino acids. Histamine,

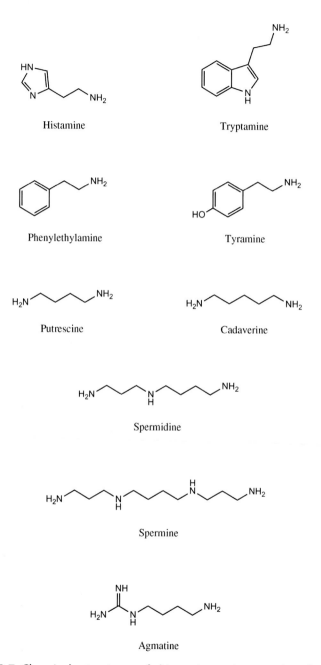

Figure 3.7 Chemical structure of biogenic amines and polyamines detected in mushrooms.

phenylethylamine, tyramine, putrescine, cadaverine, tryptamine, and agmatine are formed from histidine, phenylalanine, tyrosine, ornithine, lysine, tryptophan, and arginine, respectively. Relevant decarboxylases occur mainly in putrefactive and some lactic acid bacteria. Production of biogenic amines increases under conditions enabling intensive proteolysis releasing free amino acids. Health hazards decrease in the following order: histamine > tyramine > phenylethylamine; no serious risk is supposed for the other amines. Sensitivity to the three mentioned amines is greatly individual. A legislated limit of 200 mg kg^{-1} FM is therefore in place in the EU for histamine in sea fish.

The origin of biologically active polyamines spermidine and spermine is different. They are biosynthesized from diamine putrescine in all of eucaryotic organisms and have an array of biological roles. In particular, they participate in tumor cell growth and proliferation. Nevertheless, they do not initiate carcinogenesis. However, they are needed for intestinal growth and health and for wound healing (Kalač, 2014).

In a survey of 17 European wild species (Dadáková et al., 2009) directly after the harvest, putrescine was prevalent among biogenic amines, with content more than 100 mg 100 g^{-1} DM in several species, particularly within the family Boletaceae. Phenylethylamine followed, and tyramine and tryptamine occurred at very low levels. No histamine or cadaverine were detected. Spermidine contents were considerably higher than those of spermine, usually at levels of tens of mg 100 g^{-1} DM. Raw mushrooms therefore belong to food items with high spermidine levels. The highest spermidine levels occurred in spore-forming parts of fruiting bodies. Statistically significant effects of the year of harvest, age, and parts of fruit bodies and of the mutual interactions between the contents of phenylalanine, putrescine, and spermine were found in X. badius.

However, production of some biogenic amines can extensively increase during storage of mushrooms, which are known to be extremely perishable. This was proven in four species even during storage at 6°C (Kalač and Křížek, 1997) for putrescine and cadaverine. Information on changes in biogenic amine and polyamine content and composition during various preservative and cooking conditions is not yet available.

3.9 TRACE ELEMENTS

Since the 1970s, the topic of trace elements in mushroom chemistry has been the most extensively studied. This has been enabled by quickly developing and available analytical equipment, atomic absorption spectroscopy (AAS), and inductively coupled plasma (ICP). Such methods allowed the determination of even very low contents of numerous trace elements, including those with (until now) nonelucidated biological roles (eg, rare earth metals).

The pioneering work of the laboratory of Professor Ruth Seeger from the University of Würzburg, Germany, should be remembered. Approximately 400 original papers on various aspects of trace elements in mushrooms have been published so far. Therefore, it is impossible to cite most of them. Numerous references are cited in several reviews (Falandysz and Borovička, 2013; Kalač, 2010; Kalač and Svoboda, 2000); therefore, only some of the weightiest articles are cited. Most of the papers deal with topics determined from a nutritional point of view, such as health risks for humans, and others deal with environmental aspects, such as bioaccumulation of trace elements by fruit bodies from the underlying substrate. The period of testing mushrooms for indicators of environmental contamination with deleterious elements finished in the early 1990s, with the conclusion that no mushroom species can be considered as a credible bioindicator of pollution (Wondratschek and Röder, 1993).

3.9.1 Statutory Limits

Some countries have established statutory limits for detrimental elements in edible mushrooms. Due to high consumption of wild-growing species, limits of 3.0, 5.0, 2.0, and 10.0 mg kg^{-1}

DM for arsenic, total mercury, cadmium, and lead, respectively, were set in the Czech Republic, until it joined the EU in 2004. The values for the cultivated mushrooms were considerably lower: $1.0\,mg\ kg^{-1}$ DM for total mercury and cadmium. Moreover, the limits of 4.0, 80, 80, and $80\,mg\ kg^{-1}$ DM were formerly appointed for chromium, copper, iron, and zinc, respectively. Only the limits of 0.2 and $0.3\,mg\ kg^{-1}$ FM for cadmium and lead, respectively, are currently valid in the EU (EEC Directive 2001/22/EC).

According to the data of the World Health Organization (WHO Food Additives Series 63 and 64, Geneva 2011), provisional tolerable weekly intakes (PTWI) of 0.015, 0.0044, and $0.0016\,mg\ kg^{-1}$ body weight for arsenic, inorganic mercury, and methylmercury, respectively, were established. A previous PTWI of $0.007\,mg\ kg^{-1}$ body weight for cadmium was withdrawn and provisional tolerable monthly intake (PTMI) of $0.025\,mg\ kg^{-1}$ body weight was appointed. For lead, the former PTWI of $0.025\,mg\ kg^{-1}$ body weight was withdrawn and daily intake of $0.0012\,mg\ kg^{-1}$ body weight for adults was proposed due to the effect of lead on blood pressure.

3.9.2 The Effects of Environmental Factors on Trace Elements in Fruit Bodies

Commonly, trace elements in fruit bodies are species–dependent. However, the contents of individual trace elements vary widely even within a species. Ranges of one or even two orders of magnitude have been reported since the beginning of their determination in mushrooms. Age and size of fruit bodies are of less importance. Likewise, the proportion of atmospheric depositions seems to be limited due to the short lifetime of fruit bodies of most species. Nevertheless, the depositions may be important in ligniperdous mushrooms. In the author's opinion, element levels in fruit bodies of wild-growing mushrooms considerably elevate with the increasing age of mycelium and protracted lag between fructifications. Such a hypothesis goes along with the highest content of various heavy metals observed in *Agaricus campestris* harvested from the experimentally contaminated substrates in the first flush, whereas fruit bodies from the following flushes showed lower and descending contents.

There emerges geomycology, an interface discipline investigating metal–fungal interactions (for a review see Gadd, 2007). For instance, mushrooms growing in serpentine sites had higher contents of cadmium, chromium, and nickel than those from volcanic sites (Aloupi et al., 2012). Data on the ability of fruit bodies to accumulate various trace elements from underlying substrates have been reported in several articles. Such an ability is expressed by the bioconcentration factor (BCF), the ratio of the element content in the fruit body to its content in substrate (both in DM). Commonly, the uppermost layer, 10 cm deep, has been sampled and analyzed. Such a layer of forest substrate contains mainly organic debris. Nevertheless, mycelium of some species (eg, *S. luteus*) is associated with deeper mineral horizons. The published data on the BCF values thus should be perceived as approximate information.

The term hyperaccumulator has been used for mushrooms with metal contents at least 100-times higher than values in nonaccumulating species growing from the same substrate. For instance, Borovička et al. (2007) observed 800- to 2500-times higher silver content in *Amanita strobiliformis* than in its underlying substrate. Unfortunately, a very high BCF, up to several hundreds, has been known for cadmium and mercury in several edible mushroom species.

Elevated and even extremely high contents of numerous elements are typical in fruit bodies from polluted areas as compared with background levels in uncontaminated rural sites. Such situations have been well documented in mushrooms along motorways and roads with high traffic (particularly for lead during the period of leaded petrol usage), in the vicinity of various metal smelters that are operating or abandoned (Petkovšek and Pokorny, 2013), thermal power plants, in areas of both recent and historical metal mining (Árvay et al., 2014; Kojta et al., 2012), within towns with extensive redeployment of soil, or after the depositing of sewage sludge in meadows and forests.

The ability of some mushroom species to bioaccumulate some of the trace elements from the growing substrate can be used for the production of nutraceuticals and pharmacological materials. For instance, successful biofortification with selenium and lithium was developed. Both *A. bisporus* (Maseko et al., 2013) and *L. edodes* (Nunes et al., 2012) cultivated on substrates irrigated with sodium selenite (Na_2SeO_3) solution contained considerably higher levels of selenium, particularly in seleno-amino acids, which are thought to be bioavailable for humans. In another procedure, *Pleurotus florida* was cultivated on wheat straw from the seleniferous soils. Selenium content in fruit bodies was no less than 800-times higher than in the counterparts grown on wheat straw in areas that are not selenium-rich (Bhatia et al., 2013). Fruit bodies of *P. ostreatus* cultivated on coffee husks enriched with lithium chloride contained elevated levels of lithium, which showed greater bioaccessability as compared with lithium carbonate, which is often used as a psychiatric drug (de Assunção et al., 2012). Seeking the conditions of contaminated soil bioremediation with selected mushrooms has been another area of research.

Heavy metal bioaccumulation in both fruit bodies and mycelium of ectomycorrhizal species has been of great importance for the understanding of mycorrhizal processes. A great proportion of several metals, particularly zinc, copper, and cadmium, was found immobilized in mycorrhizas. A biological barrier is formed that reduces movement of the metals to tree tissues. Information on regulatory processes in the uptake and transfer of minerals to fruit bodies including spores has been very scarce (eg, Gramss and Voigt, 2013).

Most of the elements are distributed unevenly within a fruiting body. Commonly, the highest levels are observed in the spore-forming part of the cap, but not in spores; levels are lower in the rest of the cap and lowest in stipe.

Orczán et al. (2012) reported an interesting comparison of 22 elements, including major minerals, in 93 samples of 17 underground-growing mushroom species (particularly of genera *Elaphomyces* and *Tuber*) with 625 samples of aboveground mushrooms. While the contents of Al, B, Cu, Sr, and Ti were nearly identical in both groups, the levels of Ba, Ca, Mo, Na, and Zn were higher in the underground species. The levels of As, Cd, Co, Cr, Fe, K, Mg, Mn, Ni, P, Se, and V were higher in the aboveground fruit bodies.

Information on the individual chemical species (ie, various chemical compounds within an element) has been very limited until now. Such data are needed for the understanding of the biochemical and physiological processes within mushrooms and also for the evaluation of dietary bioaccessibility and bioavailability of both detrimental and nutritionally essential elements. Overall, vast data on trace element content in edible mushrooms have been available, although their credibility regarding human nutrition has been restricted. Nevertheless, similar data have dealt with many food items, even staples. More detailed data on bioconcentration ability, distribution within the fruit body, and chemical species are given for the main individual elements in the following section.

3.9.3 Main Trace Elements

The overall information on the usual content of main trace elements in fruit bodies of mushrooms collected from unpolluted areas and accumulating genera and species is presented in Table 3.6. Information and comprehensive available data for 13 elements are presented. All contents are expressed in milligram per kilogram of DM.

Contamination of analyzed fruit bodies with soil particles can be an important source of error in the determination of aluminum, iron, lanthanides, and various other metals, but usually not for determination of cadmium, mercury, and selenium. Such contamination is virtually unavoidable in the analysis of cup species and gilled fruit bodies growing underground.

Table 3.6 Usual content (mg kg^{-1} dry matter) of main trace elements in mushroom fruit bodies from unpolluted areas and accumulating genera and species known currently

Element	Content	Accumulators
Aluminum	50–200	*Amanita rubescens, Boletus edulis, Coprinus comatus, Leccinum scabrum, Macrolepiota procera, Xerocomus chrysenteron*
Arsenic	<1	*Laccaria amethystea, Laccaria laccata, Laccaria vinaceoavellanea, Agaricus* spp., *Lepista nuda, Lycoperdon perlatum*
Cadmium	0.5–5	*Agaricus* spp. (group *flavescentes*)
Chromium	0.5–5	*Armillariella mellea, Agaricus* spp., *Macrolepiota procera, Lactarius deliciosus, Lepista nuda, Marasmius oreades*
Cobalt	<0.5	*Ramaria largentii, Agaricus arvensis*
Copper	20–70	*Agaricus* spp., *Calvatia utriformis, Macrolepiota procera, M. rhacodes, Lycoperdon perlatum*
Iodine	0.05–0.2	*Clitocybe nebularis, Lepista nuda, Macrolepiota procera*
Iron	30–150	*Suillus variegatus, Suillus luteus, Hygrophoropsis aurantiaca*
Lead	1–5	*Macrolepiota rhacodes, Macrolepiota procera, Lycoperdon perlatum Agaricus* spp., *Lepista nuda, Leucoagaricus leucothites*
Mercury	<0.5–5	*Agaricus* spp., *Macrolepiota procera, Macrolepiota rhacodes, Lepista nuda, Calocybe gambosa*
Nickel	0.5–5	*Laccaria amethystea, Leccinum* spp., *Coprinus comatus, Armillariella mellea, Verpa conica*
Selenium	1–5	*Albatrellus pes-caprae, Boletus edulis, Boletus pinophilus, Boletus aestivalis, Xerocomus badius*
Zinc	30–150	*Suillus variegatus, Suillus luteus, Lycoperdon perlatum*

3.9.3.1 Aluminum

Aluminum does not rank among elements often determined in mushrooms. A wide range, from several tens to several hundreds mg kg^{-1} DM, was reported in several wild-growing species. The values are comparable or higher than those in numerous foods of plant origin. The data for cultivated species have been scarce. Nevertheless, aluminum intake from mushrooms seems to be limited due to usually low mushroom consumption.

3.9.3.2 Arsenic

Most reported contents of the metalloid arsenic have been less than $1\,mg\,kg^{-1}$ DM. The ability of bioaccumulation was observed in species of eight saprobic genera, namely *Agaricus, Calvatia, Collybia, Laccaria, Langermannia, Lepista, Lycoperdon,* and *Macrolepiota* (Vetter, 2004). Nevertheless, considerably higher contents were determined in accumulating species growing in sites polluted with arsenic, eg, about $150\,mg\,kg^{-1}$ DM in the most probably non-accumulating *X. badius* (Niedzielski et al., 2013). An extreme arsenic content of $1420\,mg\,kg^{-1}$ DM was found in highly accumulating *Laccaria amethystea* from a heavily polluted site.

Further information on arsenic speciation is needed because inorganic arsenic compounds, arsenites and arsenates, are toxic, whereas the organic counterparts have been less dangerous or nontoxic. A scale of considerably prevalent organic compounds such as arsenobetaine, dimethylarsinic acid, methylarsonic acid, and trimethylarsine oxide was reported. However, an inverse relation was recently observed in fresh and preserved *L. edodes*, with a mean proportion of 84% of inorganic arsenic in the total arsenic content (Llorente-Mirandes et al., 2014).

Mushrooms have been found to contain a higher proportion of nontoxic arsenobetaine than other terrestrial organisms. However, its role in mushrooms is still unknown. It has been hypothesized that arsenobetaine may play an osmolytic role in spore dispersal. Following experiments with mycelia of *A. bisporus, Sparassis crispa,* and *S. luteus* cultivated in axenic (sterile) substrates supplemented with various arsenic compounds, it is unlikely that the fungal mycelium is responsible for biosynthesis of arsenobetaine (Nearing et al., 2015). Thus, it is possible that arsenobetaine or its precursors are produced by the surrounding microbial community in the soil.

3.9.3.3 *Cadmium*

Cadmium is probably the most detrimental trace element in mushrooms. Its level in nonaccumulating species from unpolluted sites is commonly less than 2 mg kg^{-1} DM; however, contents between 2 and 5 mg kg^{-1} DM are not rare. Several species from the genus *Agaricus*, particularly those turning yellow after mechanical damage of their tissues, can contain several tens of mg kg^{-1} DM, even when from unpolluted sites. Widely consumed *B. edulis* and *Leccinum scabrum* are proven moderate accumulators. Worrisome levels, up to hundreds of mg kg^{-1} DM, have been repeatedly reported in mushrooms from heavily contaminated sites. Cadmium content is generally higher in caps than in stipes.

The information on chemical forms of cadmium in mushrooms has remained fragmentary. A phosphoglycoprotein and four low-molecular glycoproteins binding cadmium were isolated from *Agaricus macrosporus*. A high occurrence of cadmium-binding metallothionein-like proteins was reported in several tested species. Moreover, cystein-rich oligopeptides of the phytochelatin family were found in *B. edulis* growing in a polluted site (Collin-Hansen et al., 2007). Although the initial information from the early 1980s reported only low cadmium bioavailability from mushrooms, further works observed comparable or higher absorption from mushrooms than from inorganic cadmium salts. Cadmium levels in blood serum increased following mushroom consumption. Sun et al. (2012) recently reported a considerable decrease of cadmium bioaccessibility in cooked *A. subrufescens*. The fresh mushrooms showed the highest bioaccessibility at approximately 78% during in vitro biomimetic digestion in the stomach, followed by 69% during gastrointestinal digestion. However, microwaving the mushroom with water lowered the bioaccessibility to 58% and 50% during gastric and gastrointestinal procedures, respectively, and boiling showed even lower respective values of 51% and 46%. Thus, the available data on the health risk of cadmium from mushrooms have been fragmentary and further medical research is needed.

3.9.3.4 Chromium

The most frequently reported contents are less than $5 \, mg \, kg^{-1}$ DM, but the range of $5–10 \, mg \, kg^{-1}$ DM is not extraordinary. The accumulating species can contain more than $20 \, mg \, kg^{-1}$ DM. The prevalent nonaccumulating species were found to be bioexclusors with a bioaccumulation factor value less than 1 (García et al., 2013). The distribution of chromium between caps and stipes seems to be uneven, with higher levels in caps. The only available report (Figueiredo et al., 2007) is on the proportion of toxic hexavalent [Cr(VI)] chromium, with approximately 10% from total chromium. The BCFs were always less than 1 in 15 wild species. However, the BCF values for Cr(VI) were 10-times higher than those for total chromium. Therefore, it seems that mushrooms preferentially bioaccumulate Cr(VI) to nontoxic Cr(III), which could be a risk in substrates contaminated with Cr(VI).

3.9.3.5 Cobalt

Numerous reports concur that usual cobalt contents are up to $0.5 \, mg \, kg^{-1}$ DM, whereas levels more than $1.0 \, mg \, kg^{-1}$ DM are limited. No appreciable difference was found between ectomycorrhizal and saprobic species. Contents in caps seem to be somewhat higher than in stipes.

3.9.3.6 Copper

Copper contents up to $70 \, mg \, kg^{-1}$ DM are frequent in mushrooms from unpolluted sites. However, levels of approximately hundreds of $mg \, kg^{-1}$ DM do occur, particularly in accumulating species growing in sites contaminated with copper, such as in the vicinity of copper or polymetallic smelters. Copper is distributed unevenly within fruit bodies; two- to three-times the contents were reported in caps than in stipes of species belonging to Boletaceae family. The highest levels were found in the spore-forming parts.

Copper-binding metallothionein-like proteins were detected in 72% of 44 tested mushroom samples. The proportion is lower than those for proteins binding cadmium or zinc. Information on bioaccessibility and bioavailability of nutritionally essential copper from mushrooms has been lacking.

3.9.3.7 Iodine

Inorganic iodine contents in the groups of 27 wild-growing and three cultivated species (*A. bisporus*, *Pleurotus* spp., and *L. edodes*) were 0.28 ± 0.21 and 0.15 ± 0.09 mg kg^{-1} DM, respectively. Due to the wide ranges, both the groups did not differ significantly. The content is low; therefore, mushrooms provide a very limited contribution to the iodine requirement in human nutrition (Vetter, 2010).

3.9.3.8 Iron

Usual iron contents are 30–150 mg kg^{-1} DM, but levels surpassing 1000 mg kg^{-1} DM were reported in *Suillus variegatus*. Wide variations occur not only between species but also within some species among results of different laboratories. Somewhat higher contents were reported in ectomycorrhizal compared with saprobic species. Caps are richer in iron than stipes. Reported BCFs for several species are very low, only approximately 0.01–0.02. Knowledge of the nutritional bioavailability of iron from mushrooms has been lacking.

3.9.3.9 Lead

Lead has been considered the third risky mushroom trace element after cadmium and mercury. Lead contents in mushrooms from unpolluted areas are usually less than 5 mg kg^{-1} DM. Nevertheless, contents of tens and sometimes hundreds of mg kg^{-1} DM were determined even in nonaccumulating species growing in sites heavily polluted with lead, particularly in the areas polluted from both historical and operating mines and smelters. Bioaccumulation factors for lead are very low, less than 0.05 and usually only 0.01–0.02. Several accumulating species exist (Table 3.6). Lead seems to be distributed evenly within fruit bodies.

The initial data regarding the isotopic ratio of ^{206}Pb/^{207}Pb as an indicator of lead origin in fruit bodies are available (Borovička et al., 2014; Komárek et al., 2007). It seems that lead from recent air pollution is transferred to fruit bodies via the top organic horizon of underlying substrates. Unfortunately, information on lead bioavailability from mushrooms is lacking.

3.9.3.10 Mercury

Mercury has been assessed as the second most detrimental trace element after cadmium in mushrooms. Data on total mercury content have been vast and originate from many countries and laboratories. The extraordinary contribution to the topic from various sources originated from the laboratory of Professor Jerzy Falandysz, University of Gdańsk, Poland. Their data have been published in many original articles.

Mercury contents in prevalent nonaccumulating species from unpolluted areas have been less than $1\,mg\ kg^{-1}$ DM; however, levels up to $5\,mg\ kg^{-1}$ DM have not been scarce. Even higher contents have been observed in valued *B. edulis*, *Boletus pinophilus*, and *Boletus appendiculatus*. Considerably elevated levels, even in hundreds of $mg\ kg^{-1}$ DM, were determined in accumulating species (Table 3.6) harvested from heavily polluted areas, particularly near mercury smelters. Most works report a higher level of mercury in caps than in stipes. BCFs vary widely in various species, with most values being approximately tens but with some being hundreds. As reported by Nasr et al. (2012) for 27 wild-growing species, total mercury content correlated positively with total sulfur and total nitrogen contents in fruit bodies. Sulfur-containing and nitrogen-containing compounds thus imply mercury transfer from mycelia to fruit bodies. This is in accord with an older report on proteins binding mercury in several mushroom species.

Several works up until the 1990s reported only low proportions (several percent) of highly toxic methylmercury (CH_3Hg^+). The results were recently proven by Rieder et al. (2011), who determined less than 5% of methylmercury from total mercury.

Litter-decomposing species showed the highest contents of both total mercury and methylmercury. Mushrooms accumulate methylmercury from substrates with a BCF of approximately 20 and/or methylate mercuric [Hg(II)] salts (Fischer et al., 1995). Unfortunately, even for such an important toxicant, data on bioavailability from mushrooms are lacking.

3.9.3.11 Nickel

Nickel contents have been reported often. However, these contents are low, mostly up to 5 mg kg^{-1} DM, and seem to be of no toxicological risk. There exist several accumulating species (Table 3.6) with levels of approximately tens of mg kg^{-1} DM. Information on chemical species in mushrooms and on bioavailability in humans is lacking.

3.9.3.12 Selenium

A comprehensive review of different aspects of selenium in mushrooms is available (Falandysz, 2008). Selenium in mushrooms has been quantified by several analytical methods. Particularly improper determination using flame AAS or inductively coupled plasma atomic emission spectroscopy (ICP-AES) could provide excessive values and selenium contents thus could be overestimated (Falandysz, 2013). From this point of view, the following data should be taken cautiously.

The usual contents in ectomycorrhizal species are less than 2 mg kg^{-1} DM; nevertheless, "true boletes" *B. edulis*, *B. pinophilus*, and *Boletus aestivalis* commonly contain 5–20 mg kg^{-1} DM or more (eg, Costa-Silva et al., 2011). Frequent values in saprobic species are 3–5 mg kg^{-1} DM. An extremely high accumulation of approximately 200 mg kg^{-1} DM was observed in the rarely consumed *Albatrellus pes-caprae*. As mentioned, some cultivated species were proven to accumulate organic compounds of selenium from the substrate enriched with sodium selenite. A desirable level of selenium was found in *Pleurotus fossulatus* cultivated on wheat straw originated from seleniferous soil (Bhatia et al., 2014). Such substrates could be used in regions with soils rich in selenium; however, European soils are very poor and selenium is deficient in the food chain.

Selenium in mushrooms is present as prevalent amino acids selenocysteine, selenocystine, selenomethionine bound in proteins, Se-methylselenocysteine, inorganic selenites, and several unidentified selenocompounds, including selenium-containing polysaccharides. The proportions of individual compounds vary widely in various species. Such data seem to be promising, particularly in regions deficient in dietary selenium, because this essential element is recognized as an antioxidant and antagonist of toxic effects of methylmercury. Unfortunately, the bioavailability of selenium from *B. edulis* was observed to be low (Mutanen, 1986). It would be useful to again investigate the bioaccessibility and bioavailability of selenium from the main accumulating and widely consumed species.

3.9.3.13 Zinc

Usual zinc contents vary between 25 and 200 mg kg^{-1} DM, with elevated levels observed even in accumulating species (Table 3.6). Considerably higher contents were observed in mushrooms from the vicinity of a zinc smelter. It seems that insignificant differences occur between the groups of ectomycorrhizal and saprobic species. The reported BCFs are mostly less than 10. The spore-forming part of the fruit bodies is richest in zinc. Data on the bioavailability of this essential element from mushrooms are lacking.

3.9.4 Trace Elements with Limited Data

Overall information on usual contents and (until now) known genera and species accumulating some trace elements with limited data is condensed in Table 3.7. These elements, which are nutritionally nonessential or low-risk, have been commonly determined only with additional data regarding the main trace elements described in the previous section.

Table 3.7 Usual content (mg kg^{-1}dry matter) of trace elements with limited data in mushroom fruit bodies from unpolluted areas and accumulating genera and species known currently

Element	Content	Accumulators
Antimony	0.05–0.15	*Suillus* spp., *Laccaria amethystea*, *Amanita rubescens*
Barium	1–5	*Coprinus comatus*
Boron	5–15	*Marasmius wynnei, Suillus collinitus*, *Tricholoma terreum*
Bromine	1–20	*Lepista gilva, Lepista inversa*, *Amanita rubescens*
Beryllium	<0.5–0.5	
Cesium	3–12	*Suillus luteus*
Gold	<0.02	*Lycoperdon perlatum, Boletus edulis*, *Cantharellus lutescens, Morchella esculenta*
Lanthanides (15 metals)	~1 for main metals Ce > La > Nd	
Lithium	0.1–0.2	*Craterellus cornucopioides*, *Amanita strobiliformis*
Manganese	10–60	*Agaricus* spp.
Molybdenum	<0.5	
Rubidium	Tens to hundreds	Boletaceae family
Silver	0.2–3	*Amanita strobiliformis, Agaricus* spp., *Boletus edulis, Lycoperdon perlatum*
Strontium	<2	*Amanita rubescens*
Thallium	<0.25	
Thorium	<0.15	
Tin	<1	
Titanium	<10	
Uranium	<0.03	
Vanadium	<0.5	*Coprinus comatus*, toxic *Amanita muscaria*

3.9.5 Concluding Remarks

Health consequences of detrimental trace elements cadmium, mercury, lead, and arsenic consumed in wild-growing mushrooms cannot be assessed thoroughly. Our knowledge of chemical species, bioaccessibility, and bioavailability of all trace elements from mushrooms, including those essential for human nutrition, has been only fragmentary. Contents of the "total element" are therefore used for calculation of the intake.

The consideration of mushroom trace elements in human nutrition has been furthermore limited due to the very scarce data on changes in their contents during various preserving and cooking methods. For instance, the detrimental elements were leached during boiling to a higher extent from destroyed tissues of frozen mushroom slices than from fresh, air-dried, or freeze-dried slices, with the released proportion in the following order: Cd > Pb > Hg.

Overall, even in accumulating species (because these are rarely consumed alone) from unpolluted areas, the intake of the detrimental elements can scarcely surpass limits given in Section 3.9.1. However, the health risk considerably increases in mushrooms grown in polluted and chiefly in heavily polluted areas, such as in the vicinity of metal mines and smelters. Such mushrooms should not be consumed at all. However, mushrooms cultivated on unpolluted substrates are safe because levels of detrimental elements are low.

REFERENCES

Akyüz, M., Onganer, A.N., Erecevit, P., Kirbağ, S., 2012. Flavonoid contents and 2,2-diphenyl-1-picrylhydrazyl radical scavenging activity of some edible mushrooms from Turkey: *A. bisporus* and *Pleurotus* spp. Curr. Top. Nutraceutical Res. 10, 133–136.

Aloupi, M., Koutrotsios, G., Koulousaris, M., Kalogeropoulos, N., 2012. Trace metal contents in wild edible mushrooms growing on serpentine and volcanic soils on the island of Lesvos, Greece. Ecotoxicol. Environ. Saf. 78, 184–194.

Aprea, E., Romano, A., Betta, E., Biasioli, F., Cappellin, L., Fanti, M., et al., 2015. Volatile compound changes during shelf life of dried *Boletus edulis*: comparison between SPME-GC-MS and PTR-ToF-MS analysis. J. Mass Spectrom. 50, 56–64.

Árvay, J., Tomáš, J., Hauptvogl, M., Kopernická, M., Kováčik, A., Bajčan, D., et al., 2014. Contamination of wild-grown edible mushrooms by heavy metals in a former mercury-mining area. J. Environ. Sci. Health B 49, 815–827.

Ayaz, F.A., Torun, H., Özel, A., Col, M., Duran, C., Sesli, E., et al., 2011. Nutritional value of some wild edible mushrooms from the Black Sea region (Turkey). Turk. J. Biochem. 36, 385–393.

Barros, L., Dueñas, M., Ferreira, I.C.F.R., Baptista, P., Santos-Buelga, C., 2009. Phenolic acids determination by HPLC-DAD-ESI/MS in sixteen different Portuguese wild mushroom species. Food Chem. Toxicol. 47, 1076–1079.

Barros, L., Pereira, C., Ferreira, I.C.F.R., 2013. Optimized analysis of organic acids in edible mushrooms from Portugal by ultra fast liquid chromatography and photodiode array detection. Food Anal. Methods 6, 309–316.

Beara, I.N., Lesjak, M.M., Četojević-Simin, D.D., Marjanović, Ž.S., Ristić, J.D., Mrkonjić, Z.O., et al., 2014. Phenolic profile, antioxidant, anti-inflammatory and cytotoxic activities of black (*Tuber aestivum* Vittad.) and white (*Tuber magnatum* Pico) truffles. Food Chem. 165, 460–466.

Beluhan, S., Ranogajec, A., 2011. Chemical composition and non-volatile components of Croatian wild edible mushrooms. Food Chem. 124, 1076–1082.

Bhatia, P., Aureli, F., D'Amato, M., Prakash, R., Cameotra, S.S., Nagaraja, T.P., et al., 2013. Selenium bioaccessibility and speciation in biofortified *Pleurotus* mushrooms grown on selenium-rich agricultural residues. Food Chem. 140, 225–230.

Bhatia, P., Bansal, C., Prakash, R., Nagaraja, T.P., 2014. Selenium uptake and associated anti-oxidant properties in *Pleurotus fossulatus* cultivated on wheat straw from seleniferous fields. Acta Aliment. 43, 280–287.

Borovička, J., Řanda, Z., Jelínek, E., Kotrba, P., Dunn, C.E., 2007. Hyperaccumulation of silver by *Amanita strobiliformis* and related species of the section *Lepidella*. Mycol. Res. 111, 1339–1344.

Borovička, J., Mihaljevič, M., Gryndler, M., Kubrová, J., Žigová, A., Hršelová, H., et al., 2014. Lead isotopic signatures of saprotrophic macrofungi of various origin: tracing for lead sources and possible applications in geomycology. Appl. Geochem. 43, 114–120.

Chen, W., Li, W., Yang, Y., Yu, H., Zhou, S., Feng, J., et al., 2015. Analysis and evaluation of tasty components in the pileus and stipe of *Lentinula edodes* at different growth stages. J. Agric. Food Chem. 63, 795–801.

Chiang, P.D., Yen, C.T., Mau, J.L., 2006. Non-volatile taste components of canned mushrooms. Food Chem. 97, 431–437.

Collin-Hansen, C., Pedersen, S.A., Andersen, R.A., Steinnes, E., 2007. First report of phytochelatins in a mushroom: induction of phytochelatins by metal exposure in *Boletus edulis*. Mycologia 99, 161–174.

Combet, E., Henderson, J., Eastwood, D.C., Burton, K.S., 2006. Eight-carbon volatiles in mushrooms and fungi: properties, analysis, and biosynthesis. Mycoscience 47, 317–326.

Combet, E., Henderson, J., Eastwood, D.C., Burton, K.S., 2009. Influence of sporophore development, damage, storage, and tissue specificity on the enzymatic formation of volatiles in mushrooms (*Agaricus bisporus*). J. Agric. Food Chem. 57, 3709–3717.

Costa-Silva, F., Marques, G., Matos, C.C., Barros, A.I.R.N.A., Nunes, F.M., 2011. Selenium contents of Portuguese commercial and wild edible mushrooms. Food Chem. 126, 91–96.

Dadáková, E., Pelikánová, T., Kalač, P., 2009. Content of biogenic amines and polyamines in some species of European wild-growing edible mushrooms. Eur. Food Res. Technol. 230, 163–171.

De Assunção, L.S., da Luz, J.M.R., da Silva, M.C.S., Vieira, P.A.F., Bazzolli, D.M.S., Vanetti, M.C.D., et al., 2012. Enrichment of mushrooms: an interesting strategy for the acquisition of lithium. Food Chem. 134, 1123–1127.

De Pinho, P.G., Ribeiro, B., Gonçalves, R.F., Baptista, P., Valentão, P., Seabra, R.M., et al., 2008. Correlation between the pattern volatiles and the overall aroma of wild edible mushrooms. J. Agric. Food Chem. 56, 1704–1712.

Dubost, N.J., Ou, B., Beelman, R.B., 2007. Quantification of polyphenols and ergothioneine in cultivated mushrooms and correlation to total antioxidant capacity. Food Chem. 105, 727–735.

Falandysz, J., 2008. Selenium in edible mushrooms. J. Environ. Sci. Health C 26, 256–299.

Falandysz, J., 2013. Review: on published data and methods for selenium in mushrooms. Food Chem. 138, 242–250.

Falandysz, J., Borovička, J., 2013. Macro and trace mineral constituents and radionuclides in mushrooms: health benefits and risks. Appl. Microbiol. Biotechnol. 97, 477–501.

Fernandes, Â., Barros, L., Antonio, A.L., Barreira, J.C.M., Oliveira, M.B.P.P., Martins, A., et al., 2014. Using gamma irradiation to attenuate the effects caused by drying or freezing in *Macrolepiota procera* organic acids and phenolic compounds. Food Bioprocess Technol. 7, 3012–3021.

Fernandes, Â., Barreira, J.C.M., Antonio, A.L., Rafalski, A., Oliveira, M.B.P.P., Martins, A., et al., 2015. How does electron beam irradiation dose affect the chemical and antioxidant profiles of wild dried *Amanita* mushrooms? Food Chem. 182, 309–315.

Figueiredo, E., Soares, M.E., Baptista, P., Castro, M., Bastos, M.L., 2007. Validation of an electrothermal atomization atomic absorption spectrometry method for quantification of total chromium and chromium(VI) in wild mushrooms and underlying soils. J. Agric. Food Chem. 55, 7192–7198.

Fischer, R.G., Rapsomanikis, S., Andreae, M.O., Baldi, F., 1995. Bioaccumulation of methylmercury and transformation of inorganic mercury by macrofungi. Environ. Sci. Technol. 29, 993–999.

Gadd, G.M., 2007. Geomycology: biogeochemical transformation of rocks, minerals, metals and radionuclides by fungi, bioweathering and bioremediation. Mycol. Res. 111, 3–49.

García, M.A., Alonso, J., Melgar, M.J., 2013. Bioconcentration of chromium in edible mushrooms: influence of environmental and genetic factors. Food Chem. Toxicol. 58, 249–254.

Gąsecka, M., Mleczek, M., Siwulski, M., Niedzielski, P., Kozak, L., 2015. The effect of selenium on phenolics and flavonoids in selected edible white rot fungi. LWT—Food Sci. Technol. 63, 726–731.

Geösel, A., Sipos, L., Stefanovits-Bányai, É., Kókai, Z., Györfi, J., 2011. Antioxidant, polyphenol and sensory analysis of *Agaricus bisporus* and *Agaricus subrufescens* cultivars. Acta Aliment. 40 (Suppl.), 33–40.

Gil-Ramírez, A., Aldars-García, L., Palanisamy, M., Jiverdeanu, R.M., Ruiz-Rodríguez, A., Marín, F.R., et al., 2013. Sterol enriched fractions obtained from *Agaricus bisporus* fruiting bodies and by-products by compressed fluid technologies (PLE and SFE). Innovative Food Sci. Emerging Technol. 18, 101–107.

Glamočlija, J., Stojković, D., Nikolić, M., Ćirić, A., Reis, F.S., Barros, L., et al., 2015. A comparative study on edible *Agaricus* mushrooms as functional foods. Food Funct. 6, 1900–1910.

Gramss, G., Voigt, K.-D., 2013. Clues for regulatory processes in fungal uptake and transfer of minerals to the basidiospore. Biol. Trace Elem. Res. 154, 140–149.

Gross, B., Asther, M., 1989. [Aromas from Basidiomycetes: characteristics, analysis and productions.] Sci. Aliments. 9, 427–454. (in French).

Grosshauser, S., Schieberle, P., 2013. Characterization of the key odorants in pan-fried white mushrooms (*Agaricus bisporus* L.) by means of molecular sensory science: comparison with the raw mushroom tissue. J. Agric. Food Chem. 61, 3804–3813.

Guo, Y.J., Deng, G.F., Xu, X.R., Wu, S., Li, S., Xia, E.Q., et al., 2012. Antioxidant capacities, phenolic compounds and polysaccharide contents of 49 edible macro-fungi. Food Funct. 3, 1195–1205.

Heleno, S.A., Barros, L., Sousa, M.J., Martins, A., Santos-Buelga, C., Ferreira, I.C.F.R., 2011. Targeted metabolites analysis in wild *Boletus* species. LWT—Food Sci. Technol. 44, 1343–1348.

Heleno, S.A., Stojković, D., Barros, L., Glamočlija, J., Soković, M., Martins, A., et al., 2013. A comparative study of chemical composition, antioxidant and antimicrobial properties of *Morchella esculenta* (L.) Pers. from Portugal and Serbia. Food Res. Int. 51, 236–243.

Heleno, S.A., Martins, A., Queiroz, M.J.R.P., Ferreira, I.C.F.R., 2015. Bioactivity of phenolic acids: metabolites *versus* parent compounds: a review. Food Chem. 173, 501–513.

Hiraide, M., Kato, A., Nakashima, T., 2010. The smell and odorous components of dried shiitake mushroom, *Lentinula edodes*. V: changes in lenthionine and lentinic acid contents during the drying process. J. Wood Sci. 56, 477–482.

Holighaus, G., Weissbecker, B., Fragstein, M., Schütz, S., 2014. Ubiquitous eight-carbon volatiles of fungi are infochemicals for a specialist fungivore. Chemoecology 24, 57–66.

Inamdar, A.A., Hossain, M.M., Bernstein, A.I., Miller, G.W., Richardson, J.R., Bennett, J.W., 2013. Fungal-derived semiochemical 1-octen-3-ol disrupts dopamine packaging and causes neurodegeneration. Proc. Nat. Acad. Sci. USA 110, 19561–19566.

Jang, I.T., Hyun, S.H., Shin, J.W., Lee, Y.H., Ji, J.H., & Lee, J.S., 2014. Characterization of an anti-gout xanthine oxidase inhibitor from *Pleurotus ostreatus*. Mycobiology 42, 296–300.

Ji, H., Du, A., Zhang, L., Li, S., Yang, M., Li, B., 2012. Effect of drying methods on antioxidant properties and phenolic content in white button mushroom. Int. J. Food Eng. 15, 58–65.

Jolivet, S., Arpin, N., Wichers, H.J., Pellon, G., 1998. *Agaricus bisporus* browning: a review. Mycol. Res. 102, 1459–1483.

Kalač, P., 2010. Trace element contents in European species of wild growing edible mushrooms: a review for the period 2000–2009. Food Chem. 122, 2–15.

Kalač, P., 2014. Health effects and occurrence of dietary polyamines. A review for the period 2005–mid 2013. Food Chem. 161, 27–39.

Kalač, P., Křížek, M., 1997. Formation of biogenic amines in four edible mushroom species stored under different conditions. Food Chem. 58, 233–236.

Kalač, P., Svoboda, L., 2000. A review of trace element concentrations in edible mushrooms. Food Chem. 69, 273–281.

Kalaras, M.D., Beelman, R.B., Holick, M.F., Elias, R.J., 2012. Generation of potentially bioactive ergosterol-derived products following pulsed ultraviolet light exposure of mushrooms (*Agaricus bisporus*). Food Chem. 135, 396–401.

Kalogeropoulos, N., Yanni, A.E., Koutrotsios, G., Aloupi, M., 2013. Bioactive microconstituents and antioxidant properties of wild edible mushrooms from the island of Lesvos, Greece. Food Chem. Toxicol. 55, 378–385.

Kaneko, K., Kudo, Y., Yamanobe, T., Mawatari, K., Yasuda, M., Nakagomi, K., et al., 2008. Purine contents of soybean-derived foods and selected Japanese vegetables and mushrooms. Nucleosides, Nucleotides, and Nucleic Acids 27, 628–630.

Kawai, M., Sekine-Hayakawa, Y., Okiyama, A., Ninomiya, Y., 2012. Gustatory sensation of L- and D-amino acids in humans. Amino Acids 43, 2349–2358.

Kim, M.Y., Chung, I.M., Lee, S.J., Ahn, J.K., Kim, E.H., Kim, M.J., et al., 2009. Comparison of free amino acid, carbohydrates concentrations in Korean edible and medicinal mushrooms. Food Chem. 113, 386–393.

Kivrak, I., 2015. Analytical methods applied to assess chemical composition, nutritional value and in vitro bioactivities of *Terfezia olbiensis* and *Terfezia claveryi* from Turkey. Food Anal. Methods 8, 1279–1293.

Kojta, A.K., Jarzyńska, G., Falandysz, J., 2012. Mineral composition and heavy metal accumulation capacity of Bay Bolete (*Xerocomus badius*) fruiting bodies collected near a former gold and copper mining area. J. Geochem. Explor. 121, 76–82.

Kolayli, S., Sahin, H., Aliyazicioglu, R., Sesli, E., 2012. Phenolic components and antioxidant activity of three edible wild mushrooms from Trabzon, Turkey. Chem. Nat. Compd. 48, 137–140.

Komárek, M., Chrastný, V., Štíchová, J., 2007. Metal/metalloid contamination and isotopic composition of lead in edible mushrooms and forest soils originating from a smelting area. Environ. Int. 33, 677–684.

Kramer, R., Abraham, W.-R., 2012. Volatile sesquiterpenes from fungi: what are they good for? Phytochem. Rev. 11, 15–37.

Krings, U., Berger, R.G., 2014. Dynamics of sterols and fatty acids during UV-B treatment of oyster mushroom. Food Chem. 149, 10–14.

Krzyczkowski, W., Malinowska, E., Suchocki, P., Kleps, J., Olejnik, M., Herold, F., 2009. Isolation and quantitative determination of ergosterol peroxide in various edible mushroom species. Food Chem. 113, 351–355.

Lam, Y.S., Okello, E.J., 2015. Determination of lovastatin, β-glucan, total polyphenols, and antioxidant activity in raw and processed oyster culinary-medicinal mushroom, *Pleurotus ostreatus* (higher Basidiomycetes). Int. J. Med. Mushrooms 17, 117–128.

Leal, A.R., Barros, L., Barreira, J.C.M., Sousa, M.J., Martins, A., Santos-Buelga, C., et al., 2013. Portuguese wild mushrooms at the "pharma-nutrition" interface: nutritional characterization and antioxidant properties. Food Res. Int. 50, 1–9.

Lee, Y.L., Jian, S.Y., Mau, J.L., 2009. Composition and non-volatile taste components of *Hypsizigus marmoreus*. LWT—Food Sci. Technol. 42, 594–598.

Li, H., Lee, H.S., Kim, S.H., Moon, B., Lee, C., 2014a. Antioxidant and anti-inflammatory activities of methanol extracts of *Tremella fuciformis* and its major phenolic acids. J. Food Sci. 79, C460–C468.

Li, W., Gu, Z., Zhou, S., Liu, Y., Zhang, J., 2014b. Non-volatile taste components of several cultivated mushrooms. Food Chem. 143, 427–431.

Li, Y., Ishikawa, Y., Satake, T., Kitazawa, H., Qiu, X., Rungchang, S., 2014c. Effect of active modified atmosphere packaging with different initial gas compositions on nutritional compounds of shiitake mushrooms (*Lentinus edodes*). Postharvest Biol. Technol. 92, 107–113.

Li, X., Feng, T., Zhou, F., Zhou, S., Liu, Y., Li, W., et al., 2015. Effects of drying methods on the tasty compounds of *Pleurotus eryngii*. Food Chem. 166, 358–364.

Lin, J.T., Liu, C.W., Chen, Y.C., Hu, C.C., Juang, L.D., Shiesh, C.C., et al., 2014. Chemical composition, antioxidant and anti-inflammatory properties for ethanolic extracts from *Pleurotus eryngii* fruiting bodies harvested at different time. LWT—Food Sci. Technol. 55, 374–382.

Liu, Y., Huang, F., Yang, H., Ibrahim, S.A., Wang, Y., Huang, W., 2014. Effects of preservation methods on amino acids and 5′-nucleotides of *Agaricus bisporus* mushrooms. Food Chem. 149, 221–225.

Llorente-Mirandes, T., Barbero, M., Rubio, R., López-Sánchez, J.F., 2014. Occurrence of inorganic arsenic in edible Shiitake (*Lentinula edodes*) products. Food Chem. 158, 207–215.

Lou, S.N., Montag, A., 1994. [Changes in the nucleostatus of mushrooms during storage and thermal processing.] Dtsch. Lebensmitt. Rundsch. 90, 278–284. (in German).

Maga, J.A., 1981. Mushroom flavor. J. Agric. Food Chem. 29, 1–4.

Malheira, R., de Pinho, P.G., Soares, S., da Silva Ferreira, A.C., Baptista, P., 2013. Volatile biomarkers for wild mushrooms species discrimination. Food Res. Int. 54, 186–194.

Maseko, T., Callahan, D.L., Dunshea, F.R., Doronila, A., Kolev, S.D., Ng, K., 2013. Chemical characterisation and speciation of organic selenium in cultivated selenium-enriched *Agaricus bisporus*. Food Chem. 141, 3681–3687.

Mattila, P., Könkö, K., Pihlava, J.-M., Astola, J., Vahteristo, L., Hietaniemi, V., et al., 2001. Contents of vitamins, mineral elements, and some phenolic compounds in cultivated mushrooms. J. Agric. Food Chem. 49, 2343–2348.

Mau, J.L., 2005. The umami taste of edible and medicinal mushrooms. Int. J. Med. Mushrooms 7, 113–119.

Muszyńska, B., Sułkowska-Ziaja, K., 2012. Analysis of indole compounds in edible *Basidiomycota* species after thermal processing. Food Chem. 132, 455–459.

Muszyńska, B., Sułkowska-Ziaja, K., Ekicrt, H., 2009. Indole compounds in fruiting bodies of some selected *Macromycetes* species and in their mycelia cultured in vitro. Pharmazie 64, 479–480.

Muszyńska, B., Maslanka, A., Ekiert, H., Sułkowska-Ziaja, K., 2011a. Analysis of indole compounds in *Armillaria mellea* fruiting bodies. Acta Pol. Pharm. 68, 93–97.

Muszyńska, B., Sułkowska-Ziaja, K., Ekiert, H., 2011b. Indole compounds in fruiting bodies of some edible *Basidiomycota* species. Food Chem. 125, 1306–1308.

Muszyńska, B., Sułkowska-Ziaja, K., Wójcik, A., 2013a. Levels of physiologically active indole derivatives in the fruiting bodies of some edible mushrooms (*Basidiomycota*) before and after thermal processing. Mycoscience 54, 321–326.

Muszyńska, B., Sułkowska-Ziaja, K., Ekiert, H., 2013b. Analysis of indole compounds in methanolic extracts from the fruiting bodies of *Cantharellus cibarius* (the Chantarelle) and from the mycelium of this species cultured in vitro. J. Food Sci. Technol. 50, 1233–1237.

Muszyńska, B., Sułkowska-Ziaja, K., Ekiert, H., 2013c. Phenolic acids in selected edible *Basidiomycota* species: *Armillaria mellea*, *Boletus badius*, *Boletus edulis*, *Cantharellus cibarius*, *Lactarius deliciosus* and *Pleurotus ostreatus*. Acta Sci. Pol., Hortorum Cultus 12, 107–116.

Muszyńska, B., Kała, K., Sułkowska-Ziaja, K., Gawel, K., Zając, M., Opoka, W., 2015. Determination of indole compounds released from selected edible mushrooms and their biomass to artificial stomach juice. LWT—Food Sci. Technol. 62, 27–31.

Mutanen, M., 1986. Bioavailability of selenium in mushrooms, *Boletus edulis*, to young women. Int. J. Vitam. Nutr. Res. 56, 297–301.

Nasr, M., Malloch, D.W., Arp, P.A., 2012. Quantifying Hg within ectomycorrhizal fruiting bodies, from emergence to senescence. Fungal Biol. 116, 1163–1177.

Nearing, M.M., Koch, I., Reimer, K.J., 2015. Uptake and transformation of arsenic during the vegetative life stage of terrestrial fungi. Environ. Pollut. 197, 108–115.

Niedzielski, P., Mleczck, M., Magdziak, Z., Siwulski, M., Kozak, L., 2013. Selected arsenic species: As(III), As(V) and dimethylarsenic acid (DMMA) in *Xerocomus badius* fruiting bodies. Food Chem. 141, 3571–3577.

Nowacka, N., Nowak, R., Drozd, M., Olech, M., Los, R., Malm, A., 2014. Analysis of phenolic constituents, antiradical and antimicrobial activity of edible mushrooms growing in Poland. LWT—Food Sci. Technol. 59, 689–694.

Nunes, R.C.F.L., da Luz, J.M.R., Freitas, R.B., Higuchi, A., Kasuya, M.C.M., Vanetti, M.C.D., 2012. Selenium bioaccumulation in shiitake mushrooms: a nutritional alternative source of this element. J. Food Sci. 77, C983–C986.

Obodai, M., Ferreira, I.C.F.R., Fernandes, A., Barro, L., Narh Mensah, D.L., Dzomeku, M., et al., 2014. Evaluation of the chemical and antioxidant properties of wild and cultivated mushrooms of Ghana. Molecules 19, 19532–19548.

Orczán, Á.K., Vetter, J., Merényi, Z., Bonifert, É., Bratek, Z., 2012. Mineral composition of hypogeous fungi in Hungary. J. Appl. Bot. Food Qual. 85, 100–104.

Pei, F., Shi, Y., Gao, X., Wu, F., Mariga, A.M., Yang, W., et al., 2014. Changes in non-volatile taste components of button mushroom (*Agaricus bisporus*) during different stages of freeze drying and freeze drying combined with microwave vacuum drying. Food Chem. 165, 547–554.

Petkovšek, S.A.S., Pokorny, B., 2013. Lead and cadmium in mushrooms in the vicinity of two large emission sources in Slovenia. Sci. Total Environ. 443, 944–954.

Petrović, J., Stojković, D., Reis, F.S., Barros, L., Glamočlija, J., Ćirić, A., et al., 2014. Study on chemical, bioactive and food preserving properties of *Laetiporus sulphureus* (Bull: Fr.) Murr. Food Funct. 5, 1441–1451.

Radzki, W., Sławińska, A., Jablońska-Ryś, E., Gustaw, W., 2014. Antioxidant capacity and polyphenolic content of dried wild edible mushrooms from Poland. Int. J. Med. Mushrooms 16, 65–75.

Ramsden, C.A., Riley, P.A., 2014. Tyrosinase: the four oxidation states of the active site and their relevance to enzymatic activation, oxidation and inactivation. Bioorg. Med. Chem. 22, 2388–2395.

Reis, F.S., Martins, A., Barros, L., Ferreira, I.C.F.R., 2012b. Antioxidant properties and phenolic profile of the most widely appreciated cultivated mushrooms: A comparative study between *in vivo* and *in vitro* samples. Food Chem. Toxicol. 50, 1201–1207.

Reis, F.S., Barros, L., Sousa, M.J., Martins, A., Ferreira, I.C.F.R., 2014a. Analytical methods applied to the chemical characterization and antioxidant properties of three wild edible mushroom species from Northeastern Portugal. Food Anal. Methods 7, 645–652.

Reis, F.S., Stojković, D., Barros, L., Glamočlija, J., Ćirić, A., Soković, M., et al., 2014b. Can *Suillus granulatus* (L.) Roussel be classified as a functional food? Food Funct. 5, 2861–2869.

Ribeiro, B., Andrade, P.B., Silva, B.M., Baptista, P., Seabra, R.M., Valentão, P., 2008a. Comparative study on free amino acid composition of wild edible mushroom species. J. Agric. Food Chem. 56, 10973–10979.

Ribeiro, B., Lopes, R., Andrade, P.B., Seabra, R.M., Gonçalves, R.F., Baptista, P., et al., 2008b. Comparative study of phytochemicals and antioxidant potential of wild edible mushroom caps and stipes. Food Chem. 110, 47–56.

Rieder, S.R., Brunner, I., Horvat, M., Jacobs, A., Frey, B., 2011. Accumulation of mercury and methylmercury by mushrooms and earthworms from forest soils. Environ. Pollut. 159, 2861–2869.

Rotzoli, N., Dunkel, A., Hofmann, T., 2006. Quantitative studies, taste reconstitution, and omission experiments on the key taste compounds in morel mushroom (*Morchella deliciosa* Fr.). J. Agric. Food Chem. 54, 2705–2711.

Savage, G.P., Nilzen, V., Osterberg, K., Vanhanen, L., 2002. Soluble and insoluble oxalate content in mushroom. Int. J. Food Sci. Nutr. 53, 293–296.

Sembratowicz, I., Rusinek-Prystupa, E., 2012. Content of cadmium, lead, and oxalic acid in wild edible mushrooms harvested in places with different pollution levels. Pol. J. Environ. Stud. 21, 1825–1830.

Sommer, I., Schwartz, H., Solar, S., Sontag, G., 2009. Effect of γ-irradiation on agaritine, γ-glutamyl-4-hydroxybenzene (GHB), antioxidant capacity, and total phenolic content of mushrooms (*Agaricus bisporus*). J. Agric. Food Chem. 57, 5790–5794.

Sommer, I., Schwartz, H., Solar, S., Sontag, G., 2010. Effect of gamma-irradiation on flavour 5′-nucleotides, tyrosine, and phenylalanine in mushrooms (*Agaricus bisporus*). Food Chem. 123, 171–174.

Stojković, D., Reis, F.S., Barros, L., Glamočlija, J., Ćirić, A., van Griensven, L.J.L.D., et al., 2013. Nutrients and non-nutrients composition and bioactivity of wild and cultivated *Coprinus comatus* (O.F.Müll.) Pers. Food Chem. Toxicol. 59, 289–296.

Stojković, D., Reis, F.S., Glamočlija, J., Ćirić, A., Barros, L., van Griensven, L.J.L.D., et al., 2014. Cultivated strains of *Agaricus bisporus* and *A. brasiliensis*: chemical characterization and evaluation of antioxidant and antimicrobial properties for the final healthy product—natural preservatives in yoghurt. Food Funct. 5, 1602–1612.

Sułkowska-Ziaja, K., Muszyńska, B., Ekiert, H., 2014. Analysis of indole compounds from the fruiting bodies and the culture mycelia of *Sarcodon imbricatus*. Mycoscience 55, 164–167.

Sun, L., Liu, G., Yang, M., Zhuang, Y., 2012. Bioaccessibility of cadmium in fresh and cooked *Agaricus blazei* Murill assessed by *in vitro* biomimetic digestion system. Food Chem. Toxicol. 50, 1729 1733.

Tang, Y., Li, H.M., Tang, Y.J., 2012. Comparison of sterol composition between *Tuber* fermentation mycelia and natural fruiting bodies. Food Chem. 132, 1027–1213.

Tsai, S.Y., Tsai, H.L., Mau, J.L., 2008. Non-volatile taste components of *Agaricus blazei*, *Agrocybe cylindracea* and *Boletus edulis*. Food Chem. 107, 977–983.

Tsai, S.Y., Huang, S.J., Lo, S.H., Wu, T.P., Lian, P.Y., Mau, J.L., 2009. Flavour components and antioxidant properties of several cultivated mushrooms. Food Chem. 113, 578–584.

Tsai, S.Y., Mau, J.L., Huang, S.J., 2014. Enhancement of antioxidant properties and increase of content of vitamin D_2 and non-volatile components in fresh button mushroom, *Agaricus bisporus* (higher Basidiomycetes) by gamma-irradiation. Int. J. Med. Mushrooms 16, 137–147.

Usami, A., Nakaya, S., Nakahashi, H., & Miyazawa, M., 2014. Chemical composition and aroma evaluation of volatile oils from edible mushrooms (*Pleurotus salmoneostraminus* and *Pleurotus sajor-caju*). J. Oleo. Sci. 63, 1323–1332.

Vaz, J.A., Barros, L., Martins, A., Santos-Buelga, C., Vasconselos, M.H., Ferreira, I.C.F.R., 2011a. Chemical composition of wild edible mushrooms and antioxidant properties of their water soluble polysaccharidic and ethanolic fractions. Food Chem. 126, 610–616.

Vaz, J.A., Barros, L., Martins, A., Morais, J.S., Vasconselos, M.H., Ferreira, I.C.F.R., 2011b. Phenolic profile of seventeen Portuguese wild mushrooms. LWT—Food Sci. Technol. 44, 343–346.

Velíšek, J., Cejpek, K., 2011. Pigments of higher fungi: a review. Czech J. Food Sci. 29, 87–102.

Vetter, J., 2004. Arsenic content of some edible mushroom species. Eur. Food Res. Technol. 219, 71–74.

Vetter, J., 2010. Inorganic iodine contents of common, edible mushrooms. Acta Aliment. 39, 424–430.

Vidović, S.S., Mujić, I.O., Zeković, Z.P., Lepojević, Ž.D., Tumbas, V.T., Mujić, A.I., 2010. Antioxidant properties of selected *Boletus* mushrooms. Food Biophys. 5, 49–58.

Villares, A., García-Lafuente, A., Guillamón, E., Ramos, Á., 2012. Identification and quantification of ergosterol and phenolic compounds occurring in *Tuber* spp. truffles. J. Food Compost. Anal. 26, 177–182.

Wang, S., Marcone, M.F., 2011. The biochemistry and biological properties of the world's most expensive underground edible mushroom: truffles. Food Res. Int. 44, 2567–2581.

Weijn, A., van den Berg-Somhorst, D.B.P.M., Slootweg, J.C., Vinckem, J.-P., Gruppen, H., Wichers, H.J., et al., 2013. Main phenolic compounds of the melatonin biosynthesis pathway in bruising-tolerant and bruising-sensitive button mushroom (*Agaricus bisporus*) strains. J. Agric. Food Chem. 61, 8224–8231.

Witkowska, A.M., Zujko, M.E., Mirończuk-Chodakowska, I., 2011. Comparative study of wild edible mushrooms as sources of antioxidants. Int. J. Med. Mushrooms 13, 335–341.

Woldegiorgis, A.Z., Abate, D., Haki, G.D., Ziegler, G.R., 2014. Antioxidant property of edible mushrooms collected from Ethiopia. Food Chem. 157, 30–36.

Wondratschek, I., Röder, U., 1993. Monitoring of heavy metals in soils by higher fungi. In: Markert, B. (Ed.), Plants as Biomonitors. Indicators for Heavy Metals in the Terrestrial Environment VCH, Weinheim, pp. 345–363.

Wu, F., Tang, J., Pei, F., Wang, S., Chen, G., Hu, Q., et al., 2015. The influence of four drying methods on nonvolatile taste components of White *Hypsizygus marmoreus*. Eur. Food Res. Technol. 240, 823–830.

Yamaguchi, S., Yoshikawa, T., Ikeda, S., Ninomiya, T., 1971. Measurement of the relative taste intensity of some α-amino acids and 5′-nucleotides. J. Food Sci. 36, 846–849.

Yang, J.H., Lin, H.C., Mau, J.L., 2001. Non-volatile taste components of several commercial mushrooms. Food Chem. 72, 465–471.

Yaoita, Y., Kikuchi, M., Machida, K., 2014. Terpenoids and sterols from some Japanese mushrooms. Nat. Prod. Commun. 9, 419–426.

Yildiz, O., Can, Z., Laghari, A.Q., Şahin, H., Malkoç, M., 2015. Wild edible mushrooms as a source of phenolics and antioxidants. J. Food Biochem. 39, 148–154.

Yuan, J.P., Zhao, S.Y., Wang, J.H., Kuang, H.C., Liu, X., 2008. Distribution of nucleosides and nucleobases in edible fungi. J. Agric. Food Chem. 56, 809–815.

Zawirska-Wojtasiak, R., 2004. Optical purity of (R)-(-)-1-octen-3-ol in the aroma of various species of edible mushrooms. Food Chem. 86, 113–118.

Zhang, Y., Venkitasamy, C., Pan, Z., Wang, W., 2013. Recent developments on umami ingredients of edible mushrooms—A review. Trends Food Sci. Technol. 33, 78–92.

Zhou, J., Feng, T., Ye, R., 2015. Differentiation of eight commercial mushrooms by electronic nose and gas chromatography-mass spectrometry. J. Sens.374013 http://dx.doi.org/10.1155/2015/374013.

Zhou, Z.Y., Liu, J.K., 2010. Pigments of fungi (macromycetes). Nat. Prod. Rep. 27, 1531–1570.

CHAPTER 4

Health-Stimulating Compounds and Effects

Contents

Even though this book deals primarily with nutritional views on culinary mushroom composition, some aspects of the expanding knowledge on health-stimulating compounds should be mentioned. In addition to the extensively studied medicinal mushrooms that have been used in East Asia for many centuries, the interest of researchers has recently focused on common culinary mushrooms, including wild-growing species. It is supposed that mushrooms comprise vast and mostly undiscovered pharmaceutically effective compounds. For detailed information, recent comprehensive overviews dealing with various aspects are available. Overall data on the roles of edible mushrooms in health have been presented by Roupas et al. (2012), data on antitumor potential have been presented by Ferreira et al. (2010) and Popović et al. (2013), data on cardiovascular diseases have been presented by Guillamón et al. (2010) and Choi et al. (2012), data on hypertension have been presented by Yahaya et al. (2014), data on immunity have been presented by El Enshasy and Hatti-Kaul (2013) and Guo et al. (2012), data on antiinflammatory effects have been presented by Elsayed et al. (2014), data on metabolic syndrome have been presented by Kundaković and Kolundžić (2013), and data on antibacterial applications have been presented by Schwan (2012).

Only the basic information on some health-stimulating compounds and effects are briefly provided in this chapter.

4.1 ANTIOXIDANTS

Free radicals, mainly in the form of reactive oxygen species and to a limited extent in the form of reactive nitrogen species, are produced in the normal metabolism of aerobic cells. Most of the free radicals are neutralized by cellular antioxidant defense mechanisms. The maintenance of the equilibrium between the free radicals and the antioxidant defenses is essential for the normal function of an aerobic organism. An infringed balance is known as oxidative stress. The excess of free radicals may oxidize and damage cellular DNA, proteins, and lipids. The overproduction of free radicals has been related to many disorders, including the most grave diseases, and aging. Many such diseases can be prevented. Therefore, diets with high level of natural antioxidants and free radical scavengers are recommended. Fruits and vegetables, including mushrooms are commonly known sources of such components.

A thorough review of the extensive literature on antioxidants in wild-growing mushroom species has been performed by Ferreira et al. (2009). The evaluation of the topic is not easy. The antioxidant activity of various food items, including mushrooms, has been quantified by various parameters, particularly free radicals scavenging activity, power reduction, metal chelating effects, inhibition of lipid peroxidation, and the identification of antioxidant compounds. Different results are obtained depending on the chemical nature of the determined segment(s) of antioxidants.

Moreover, the activity is not often expressed per fresh or dry matter (DM) of a mushroom, but rather in mushroom extracts with solvents of various polarity.

The antioxidants in mushrooms are mainly phenolic acids and flavonoids (see Section 3.4), followed by tocopherols (Section 2.6.1), ascorbic acid (Section 2.6.2), carotenoids (Section 2.6.1), and ergothioneine (Section 4.8). Phenolic compounds have specific health effects. They might provide benefits associated with a reduced risk of degenerative malfunctions, such as cardiovascular disease or some types of cancer. The effects are due to their ability to reduce oxidative agents by donating hydrogen from phenolic groups and quenching singlet oxygen, a widely occurring free radical. Tocopherols (vitamin E) disrupt the cascade of oxidative processes by their reaction with peroxyl radicals produced from polyunsaturated fatty acids in membrane phospholipids or lipoproteins. Ascorbic acid (vitamin C) is effective against several free radicals and protects biomembranes from lipid peroxidation damage. Cooperative interactions do exist among water-soluble ascorbic acid and fat-soluble tocopherols. Because of their limited occurrence, carotenoids in mushrooms provide low antioxidative contributions but have a very important role in plants. The information on the antioxidant activity of mushroom ergothioneine has been fragmentary. Comparable with numerous fruits and vegetables, the prevailing antioxidant activity of many mushroom species is ascribed to phenolics.

Information regarding the effect of the fruit body developmental stage on the level of antioxidants has been very incomplete. There have been reported minor differences between young and mature fruit bodies of *Agaricus subrufescens* (Soares et al., 2009), and decreased levels in mature *Lactarius deliciosus* and *Lactarius piperatus* have also been reported (Barros et al., 2007b). Sun-drying of *Agaricus bisporus* caused the highest loss of antioxidant activity to the greatest extent, followed by hot air drying and freeze drying, whereas microwave-vacuum drying was the most preserving method (Ji et al., 2012). The antioxidants considerably decreased during mushroom cooking, probably because of a destruction of polyphenols (Barros et al., 2007a). Boiling caused a higher loss of water-soluble antioxidants than frying, grilling, or microwave heating in three mushroom species (Soler-Rivas et al., 2009). On the contrary, autoclaving increased both content of polyphenolics and antioxidant activity as compared with raw fruit bodies of *Lentinula edodes* (Choi et al., 2006). Thus, the available results have been ambiguous and further research is necessary.

In a study using an in vitro simulator (Vamanu et al., 2013), several antioxidants, particularly gentisic acid and homogentisic acid from the group of phenolic acids, were released during digestion from four widely consumed mushroom species (*A. bisporus*, *Pleurotus ostreatus*, *Boletus edulis*, and *Cantharellus cibarius*). The antioxidants in synergy with mushroom fiber increased the number of propitious lactobacilli and bifidobacteria strains in the colon and the fermentative capacity of colon microflora. The wild-growing species were more effective than the cultivated ones.

4.2 BETA-GLUCANS

Hot water-soluble fractions from many medicinal and culinary–medicinal mushrooms, such as *Ganoderma lucidum*, *L. edodes*, or *Inonotus obliquus*, have been historically used as medicine in China, Japan, Korea, and eastern Russia. The active substances of such decoctions are mostly polysaccharides of various chemical compositions. Most of them belong to the group of β-glucans exhibiting antitumor and immunostimulating properties, which have units of β-D-glucopyranose connected with β-(1→3) linkages in the main chain of the glucan and additional β-(1→6) branch points. The branching structure of β-glucans plays a crucial role in their physiological activities (Bae et al., 2013). Moreover, antitumor activities were also observed in other mushroom polysaccharides and polysaccharide–protein complexes.

Such noticeably propitious health effects have naturally initiated great interest from researchers. Hundreds of original works have been published, and numerous articles have been evaluated in reviews by Mizuno and Nishitani (2013), Ren et al. (2012), Rob et al. (2009), Ruthes et al. (2015), Wasser (2002), and Zhang et al. (2011).

High-molecular-weight β-glucans are more effective than low-molecular-weight ones. Mushroom polysaccharides prevent oncogenesis and tumor metastasis. They do not attack cancer cells directly, but rather activate immune responses in the host. Their activity is particularly beneficial in clinics if used in conjunction with chemotherapy. Various β-glucans have common names (Table 4.1). Some of them are available as commercial medicines, both in their natural form and chemically modified to improve their efficacy. They are isolated from three sources – fruit bodies, cultured mycelium, or culture broth.

Beta-glucans occur in fruit bodies at level from hundreds to thousands of mg $100\,g^{-1}$ DM. Zied et al. (2014) determined in cultivated *A. subrufescens* a mean β-glucan content of approximately $5000\,mg$ $100\,g^{-1}$ DM, with a range of $2500–9700\,mg$ $100\,g^{-1}$ DM; several agronomic factors affected these results. Two of these factors, mushroom strain and casing layer, were the most important.

Table 4.1 Common name (or symbol) of β-glucans isolated from various mushroom species

Name	Species
AAG	*Auricularia auricula-judae*
Flammulin	*Flammulina velutipes*
Gl-1	*Ganoderma lucidum*
Grifolan; grifron-D	*Grifola frondosa*
Krestin (PSK polysaccharide–protein complex)	*Trametes versicolor*
Lentinan	*Lentinula edodes*
Pleuran	*Pleurotus* spp.
Schizophyllan	*Schizophyllum commune*

4.3 CARBOHYDRATES AS A POTENTIAL SOURCE OF PREBIOTICS

Prebiotics are functional food ingredients that are able to manipulate the composition of colonic microbiota in the human gut by the inhibition of exogenous pathogens and by favoring beneficial bifidobacteria and lactobacilli, thus improving the health of the host. In a simplified way, prebiotics are dietary ingredients indigestible in the upper part of the gut and are selectively fermentable by potentially favorable bacteria in the colon. Such processes produce health-promoting compounds and conditions, resulting in enhanced immune function, improved colon integrity, decreased incidence and duration of intestinal infections, reduced allergic responses, and improved feces elimination. Natural polysaccharide inulin and its derivatives and nondigestible oligosaccharides, particularly galactose oligosaccharides, have been recently added to various food items.

A candidate for use as a prebiotic has to fulfill several criteria:
- stability during food processing and storage
- resistance to the conditions in the upper gut tract (gastric acidity, hydrolysis by human digestive enzymes, and gastrointestinal absorption)
- selective fermentation in the colon to beneficial products
- selective stimulation of probiotics, such as living bifidobacteria and lactobacilli.

Potential prebiotics are mostly carbohydrates in nature. However, not every dietary carbohydrate is a prebiotic. Bioactivity of water–insoluble chitin, which accounts up to 80–90% of mushroom cell wall dry matter, is limited as compared to water-soluble polysaccharides. However, chitin is a substantial part of mushroom dietary fiber.

Among various mushroom carbohydrates, β-glucans seem to be the most promising candidates for prebiotics. They are resistant to acid hydrolysis in the stomach. Digestive enzymes secreted by the human pancreas or brush border are not able to hydrolyze β-glucosidic bonds. This makes them indigestible. Nevertheless, isolated branched β-1,3-1,6-glucan and linear α-1,3-glucan from stipes of *P. ostreatus* and *Pleurotus eryngii* tested as prebiotics showed synergic effects in a synbiotic construction with selected strains of probiotics (Synytsya et al., 2009). More details on this topic are available in an overview by Aida et al. (2009).

4.4 PROTEINS WITH SPECIFIC BIOLOGICAL ROLES

In addition to nutritionally considered proteins (Section 2.2), mushrooms contain minor proteins with specific biological effects. Among them, and of greatest interest, are lectins and hemolysins because of their pharmacological and biotechnological potential, even though some of them have known harmful effects. These related proteins differ in their effects on red blood cells in vitro. Lectins (formerly hemagglutinins) agglutinate them and hemolysins lyse them.

Moreover, mushrooms produce other proteins and peptides with interesting biological activities, such as fungal immuno-modulatory proteins, ribosome inactivating proteins, antimicrobial proteins, and enzymes such as laccases or ribonucleases. Comprehensive reviews on this topic are available (Erjavec et al., 2012; Xu et al., 2011).

Information on mushroom proteins with specific effects, both positive and negative, has been very limited regarding their thermal stability during various cooking treatments, stability in gastric juice, and particularly their absorption in the digestive tract.

4.4.1 Lectins

Lectins are ubiquitous and versatile proteins of nonimmune origin that bind reversibly and specifically to carbohydrates. Their multivalent structure enables their well-known ability to agglutinate cells. Moreover, such a specific structure presents great therapeutic and biotechnological potential. Several comprehensive reviews on mushroom lectins with numerous references are available (Guillot and Konska, 1997; Singh et al., 2010, 2015; Varrot et al., 2013; Hassan et al., 2015). Lectins participate in several physiological processes of mushrooms such as defense compounds against various pests.

Most mushroom lectins display no dietary toxicity and are not involved in mushroom poisoning. Bolesatin from toxic *Boletus satanas* and a lectin from edible *Agrocybe aegerita* (syn. *Agrocybe cylindracea*) (Jin et al., 2014) are representatives of a limited number of toxic lectins. Lectins widely occur in genera *Lactarius*, *Russula*, and *Boletus*. Some species contain several related lectins. Somewhat higher lectin content has been found in fully developed rather than in young fruit bodies.

Some mushroom lectins have been tested and used as laboratory diagnostic agents. They have also been investigated as promising antiproliferative, antitumor, antiviral, and immunity-stimulating compounds. Nevertheless, knowledge of lectins from edible mushroom species, particularly of the structural and functional characteristics, has been limited until now.

4.4.2 Hemolysins

Overall information is available from a recent review by Nayak et al. (2013). Hemolysins have been classically defined as exotoxins that are capable of lysing red blood cells. Current knowledge suggests that hemolysins are pore-forming toxins that interact with specific ligands on the surface of various target cells. They were first reported in mushrooms as early as in 1907. Until now, they were isolated and characterized in several edible and toxic species, such as *P. ostreatus* (ostreolysin and pleurotolysin), *P. eryngii* (erylysin and eryngeolysin), *Amanita rubescens* (rubescenlysin), *Flammulina velutipes* (flammutoxin), and *Schizophyllum commune* (schizolysin). Hemolysins probably participate in mushroom biology, such as in the initiation of fruit bodies or as natural insecticides.

Mushroom hemolysins have rarely been responsible for accidental poisoning because they are heat-sensitive; the poison is deactivated during thermal processing.

Recently, several mushroom hemolysins have been investigated as antitumor or antiretroviral agents, or as compounds cytotoxic to tumor cells. However, the research is in its initial stage.

4.5 LOVASTATIN

Fungal polyketide lovastatin (mevinolin) is one of the seven statins and biochemical inhibitors of 3-hydroxy-3-methylglutaryl coenzyme A reductase. Lovastatin inhibits the enzyme in the production of cholesterol and reduces the risk of coronary heart disease. Statins have been shown to have effective antiinflammatory, antioxidant, profibrinolytic, and other properties and have been widely used as drugs.

Data on lovastatin content are given in Table 4.2. There are wide variations among and within species despite the data originating

Table 4.2 Content of lovastatin (mg 100 g^{-1} dry matter) in selected mushroom species

Species	Lovastatin	Reference
Agaricus bisporus	56.5	Chen et al. (2012)
Agaricus subrufescens	18.4	Lo et al. (2012)
	0.96	Cohen et al. (2014)
Agrocybe cylindracea	58.3	Lo et al. (2012)
Auricularia polytricha	1.61	Lo et al. (2012)
Boletus edulis	32.7	Chen et al. (2012)
Coprinus comatus	0.96	Cohen et al. (2014)
Flammulina velutipes	9.08	Chen et al. (2012)
	ND	Cohen et al. (2014)
Grifola frondosa	ND	Chen et al. (2012)
	ND	Cohen et al. (2014)
Hericium erinaceus	1.44	Cohen et al. (2014)
Hypsizigus marmoreus	62.8	Lo et al. (2012)
	25.8	Chen et al. (2012)
Lentinula edodes	31.7	Lo et al. (2012)
	ND	Chen et al. (2012)
	ND	Cohen et al. (2014)
Pholiota nameko	18.6	Chen et al. (2012)
Pleurotus eryngii	15.2	Chen et al. (2012)
Pleurotus ostreatus (Japan)	60.7	Chen et al. (2012)
(South Korea)	16.5	Chen et al. (2012)
(Taiwan)	21.6	Chen et al. (2012)
(Israel)	ND	Cohen et al. (2014)
Tremella fuciformis	3.01	Lo et al. (2012)
Volvariella volvacea	5.94	Lo et al. (2012)

ND, content below detection limit.

from the same laboratory. Several analyzed species showed lovastatin content more than $50 \, mg \, 100 \, g^{-1}$ DM; nevertheless, this promising information should be supported with further data. Moreover, the information on lovastatin bioavailability from mushrooms has been lacking.

4.6 ERITADENINE

As shown in both animal and human studies, *L. edodes* possesses the ability to reduce blood cholesterol and regulate lipid metabolism (Yang et al., 2013). The active agent is a purine alkaloid eritadenine, formerly lentinacin, called 2-R,3-R-dihydroxy-4-(9-adenyl)-butyric acid (Fig. 4.1). Unlike the statins (see Section 4.5), eritadenine does not inhibit cholesterol biosynthesis in the liver, but it does enhance the removal of blood cholesterol. A diet containing 0.005% eritadenine lowered total cholesterol by 25% in serum of experimental rats.

Earlier works reported eritadenine contents in the caps of *L. edodes* to be $50–70 \, mg \, 100 \, g^{-1}$ DM, and somewhat lower levels were found in the stipes. However, Enman et al. (2007) determined $317–633 \, mg \, 100 \, g^{-1}$ DM in methanolic extracts (ie, up to 10-times higher contents). Zhang et al. (2013) observed even higher contents in ethanolic extracts of freeze-dried *L. edodes*, 1660 and $1425 \, mg \, 100 \, g^{-1}$ DM in caps and stipes, respectively. Shade drying and hot air drying lowered eritadenine contents by

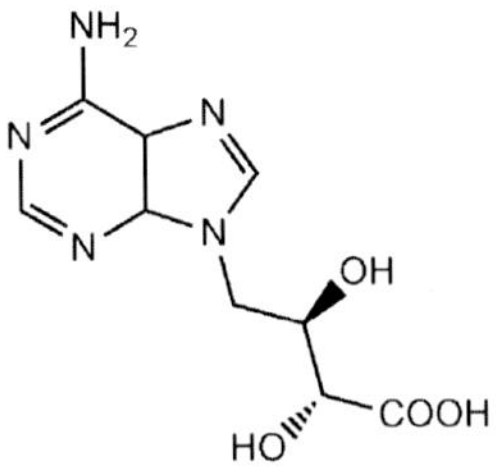

Figure 4.1 Chemical structure of eritadenine.

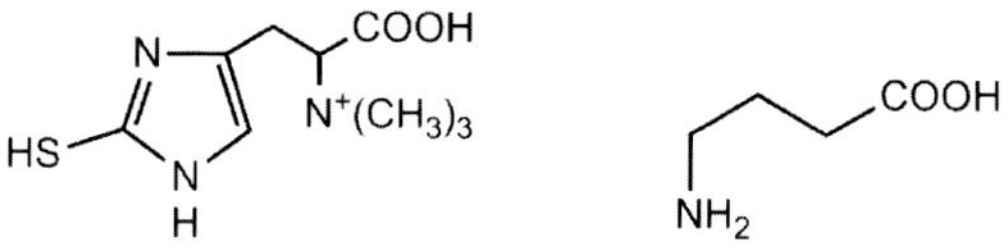

Ergothioneine   γ–Aminobutyric acid

Figure 4.2 Chemical structure of ergothioneine and gamma-aminobutyric acid (GABA).

approximately half. The losses decreased with the increasing drying temperature. It is not yet clear how changes in eritadenine during drying may affect its bioactivities. There was no clear correlation between the content of eritadenine and antioxidant activities of dried *L. edodes*.

4.7 GAMMA-AMINOBUTYRIC ACID

Dietary free amino acid γ-aminobutyric acid (GABA; Fig. 4.2) is considered a hypotensive compound. It also has an immune-amplifying role in neuroinflammatory diseases such as multiple sclerosis. Data on GABA contents are given in Table 4.3. There are apparent differences up to two orders of magnitude among species. However, in species with more available data (eg, *L. edodes*), great differences do exist even though these data originate from the same laboratory. Therefore, more data are necessary.

Although GABA contents in fruit bodies are at a level of mg per 100 g in both dry and fresh matter, Chiang et al. (2006) observed lower levels in canned *A. bisporus*, *Volvariella volvacea*, and *F. velutipes* and the corresponding broth. It seems that the blanching and thermal processing decrease GABA contents.

4.8 ERGOTHIONEINE

Water-soluble amino acid L-ergothioneine (2-mercapto-L-histidine betaine; Fig. 4.2) occurs primarily in mushrooms. No ergothioneine synthesis has been detected so far in higher plants

Table 4.3 Content of γ-aminobutyric acid (GABA) and ergothioneine (mg 100 g^{-1} dry matter) in selected mushroom species

Species	GABA	Ergothioneine	Reference
Agaricus bisporus white	–	21.0	Dubost et al. (2007)
brown	–	45.0	Dubost et al. (2007)
unspecified	12.5	93.3	Chen et al. (2012)
Agaricus campestris	–	99.0	Woldegiorgis et al. (2014)
Agaricus subrufescens	36.0	–	Tsai et al. (2008)
	184.5	3.90	Lo et al. (2012)
	39.4	3.74	Cohen et al. (2014)
Agrocybe cylindracea	21.0	–	Tsai et al. (2008)
	73.0	25.0	Lo et al. (2012)
Auricularia polytricha	28.2	0.14	Lo et al. (2012)
Boletus edulis	–	52.8 (FM)	Ey et al. (2007)
	11.0	–	Tsai et al. (2008)
	20.2	49.4	Chen et al. (2012)
Coprinus comatus	109.2	76.4	Cohen et al. (2014)
Flammulina velutipes	23.0	45.5	Chen et al. (2012)
	26.0	9.9	Cohen et al. (2014)
Grifola frondosa	–	113	Dubost et al. (2007)
	1.78	55.3	Chen et al. (2012)
	3.75	20.7	Cohen et al. (2014)
Hericium erinaceus	4.29	63.0	Cohen et al. (2014)
Hypsizigus	3.37	4.61	Lo et al. (2012)
marmoreus	11.4	41.0	Chen et al. (2012)
Laetiporus sulphureus	–	8.0	Woldegiorgis et al. (2014)
Lentinula edodes	–	198	Dubost et al. (2007)
	62.2	1.22	Lo et al. (2012)
	1.54	41.2	Chen et al. (2012)
	–	132	Woldegiorgis et al. (2014)
	18.6	33.4	Cohen et al. (2014)
Pholiota nameko	0.82	22.9	Chen et al. (2012)
Pleurotus eryngii	ND	62.5	Chen et al. (2012)
Pleurotus ostreatus (USA)	–	259	Dubost et al. (2007)
(Germany)	–	11.9 (FM)	Ey et al. (2007)
(Japan)	0.61	94.4	Chen et al. (2012)
(South Korea)	2.36	182.9	Chen et al. (2012)
(Taiwan)	ND	145.8	Chen et al. (2012)
(Ethiopia)	–	378	Woldegiorgis et al. (2014)
(Israel)	130.5	244.4	Cohen et al. (2014)

(*Continued*)

Table 4.3 Content of γ-aminobutyric acid (GABA) and ergothioneine (mg 100 g^{-1} dry matter) in selected mushroom species (Continued)

Species	GABA	Ergothioneine	Reference
Termitomyces clypeatus	–	122	Woldegiorgis et al. (2014)
Tremella fuciformis	68.7	0.73	Lo et al. (2012)
Volvariella volvacea	99.9	53.7	Lo et al. (2012)

ND, content below detection limit. FM, mg 100 g^{-1} fresh matter.

and animals, including humans. It is chiefly known as an effective antioxidant through multiple mechanisms. Moreover, it has recently attracted awareness as a biogenic key substrate of the organic cation transporter OCTN1, which is connected with autoimmune disorders. Dietary ergothioneine is taken up and concentrated mainly in mammalian mitochondria.

Data on ergothioneine content in mushrooms are presented in Table 4.3. Contents of tens of mg 100 g^{-1} DM are frequent; however, wide variations occur both among and within species. Widely consumed *P. ostreatus* and *L. edodes* seem to be rich sources of ergothioneine. In powdered samples of UV-B–irradiated mushrooms, Sapozhnikova et al. (2014) reported ergothioneine contents of 370–480, 810–920, 720–790, and 970–1040 mg 100 g^{-1} DM in brown *A. bisporus*, white *A. bisporus*, *L. edodes*, and *P. ostreatus*, respectively. The UV treatment had only little effect on ergothioneine content, except for *L. edodes*, in which a significant decrease has been observed as compared with the nonirradiated variant.

In *F. velutipes*, both refrigerated storage and boiling in water caused significant losses of ergothioneine. However, a great proportion was present in the broth (Nguyen et al., 2012). A study by Weigand-Heller et al. (2012) demonstrated that ergothioneine from dried *A. bisporus* was taken up by red blood cells postprandially and therefore was bioavailable. Mushrooms seem to serve as a worthwhile source of ergothioneine.

REFERENCES

Aida, F.M.N.A., Shuhaimi, M., Yazid, M., Maaruf, A.G., 2009. Mushroom as a potential source of prebiotics: a review. Trends Food Sci. Technol. 20, 567–575.

Bae, I.Y., Kim, H.W., Yoo, H.J., Kim, E.S., Lee, S., Park, C.Y., et al., 2013. Correlation of branching structure of mushroom β-glucan with its physiological activities. Food Res. Int. 51, 195–200.

Barros, L., Baptista, P., Correia, D.M., Morais, J.S., Ferreira, I.C.F.R., 2007a. Effects of conservation treatment and cooking on the chemical composition and antioxidant activity of Portuguese wild edible mushrooms. J. Agric. Food Chem. 55, 4781–4788.

Barros, L., Baptista, P., Estevinho, L.M., Ferreira, I.C.F.R., 2007b. Effect of fruiting body maturity stage on chemical composition and antimicrobial activity of *Lactarius* sp. mushrooms. J. Agric. Food Chem. 55, 8766–8771.

Chen, S.Y., Ho, K.J., Hsieh, Y.J., Wang, L.T., Mau, J.L., 2012. Contents of lovastatin, γ-aminobutyric acid and ergothioneine in mushroom fruiting bodies and mycelia. LWT—Food Sci. Technol. 47, 274–278.

Chiang, P.D., Yen, C.T., Mau, J.L., 2006. Non-volatile taste components of canned mushrooms. Food Chem. 97, 431–437.

Choi, Y., Lee, S.M., Chun, J., Lee, H.B., Lee, J., 2006. Influence of heat treatment on the antioxidant activities and polyphenolic compounds of shiitake (*Lentinus edodes*) mushroom. Food Chem. 99, 381–387.

Choi, E., Ham, O., Lee, S.Y., Song, B.W., Cha, M.J., Lee, C.Y., et al., 2012. Mushrooms and cardiovascular disease. Curr. Top. Nutraceutical Res. 10, 43–52.

Cohen, N., Cohen, J., Asatiani, M.D., Varshney, V.K., Yu, H.T., Yang, Y.C., et al., 2014. Chemical composition and nutritional and medicinal value of fruit bodies and submerged cultured mycelia of culinary-medicinal higher Basidiomycetes mushrooms. Int. J. Med. Mushrooms 16, 273–291.

Dubost, N.J., Ou, B., Beelman, R.B., 2007. Quantification of polyphenols and ergothioneine in cultivated mushrooms and correlation to total antioxidant capacity. Food Chem. 105, 727–735.

El Enshasy, H.A., Hatti-Kaul, R., 2013. Mushroom immunomodulators: unique molecules with unlimited applications. Trends Biotechnol. 31, 668–677.

Elsayed, E.A., El Enshasy, H., Wadaan, M.A.M., Aziz, R., 2014. Mushrooms: a potential natural source of anti-inflammatory compounds for medical applications. Mediators Inflamm. http://dx.doi.org/10.1155/2014/805841.

Enman, J., Rova, U., Berglund, K.A., 2007. Quantification of the bioactive compound eritadenine in selected strains of shiitake mushroom (*Lentinus edodes*). J. Agric. Food Chem. 55, 1177–1180.

Erjavec, J., Kos, J., Ravnikar, M., Dreo, T., Sabotič, J., 2012. Proteins of higher fungi—from forest to application. Trends Biotechnol. 30, 259–273.

Ey, J., Schömig, E., Taubert, D., 2007. Dietary sources and antioxidant effects of ergothioneine. J. Agric. Food Chem. 55, 6466–6474.

Ferreira, I.C.F.R., Barros, L., Abreu, R.M.V., 2009. Antioxidants in wild mushrooms. Curr. Med. Chem. 16, 1543–1560.

Ferreira, I.C.F.R., Vaz, J.A., Vasconcelos, M.H., Martins, A., 2010. Compounds from wild mushrooms with antitumor potential. Anticancer Agents Med. Chem. 10, 424–436.

Guillamón, E., García-Lafuente, A., Lozano, M., D'Arrigo, M., Rostagno, M.A., Villares, A., et al., 2010. Edible mushrooms: role in the prevention of cardiovascular diseases. Fitoterapia 81, 715–723.

Guillot, J., Konska, G., 1997. Lectins in higher fungi. Biochem. Syst. Ecol. 25, 203–230.

Guo, C., Choi, M.W., Cheung, P.C.K., 2012. Mushroom and immunity. Curr. Top. Nutraceutical Res. 10, 31–41.

Hassan, M.A.A., Rouf, R., Tiralongo, E., May, T.W., Tiralongo, J., 2015. Mushroom lectins: specificity, structure and bioactivity relevant to human disease. Int. J. Mol. Sci. 16, 7802–7838.

Ji, H., Du, A., Zhang, L., Li, S., Yang, M., Li, B., 2012. Effect of drying methods on antioxidant properties and phenolic content in white button mushroom. Int. J. Food Eng. 15, 58–65.

Jin, Y.X., et al., 2014. Lethal protein in mass consumption edible mushroom *Agrocybe aegerita* linked to strong hepatic toxicity. Toxicon. http://dx.doi.org/10.1016/toxicon.2014.08.066.

Kundaković, T., Kolundžić, M., 2013. Therapeutic properties of mushrooms in managing adverse effects in the metabolic syndrome. Curr. Top. Med. Chem. 13, 2734–2744.

Lo, Y.C., Lin, S.Y., Ulziijargal, E., Chen, A.Y., Chien, R.C., Tzou, Y.J., et al., 2012. Comparative study of contents of several bioactive components in fruiting bodies and mycelia of culinary-medicinal mushrooms. Int. J. Med. Mushrooms 14, 357–363.

Mizuno, M., Nishitani, Y., 2013. Immunomodulating compounds in Basidiomycetes. J. Clin. Biochem. Nutr. 52, 202–207.

Nayak, A.P., Green, B.J., Beezhold, D.H., 2013. Fungal hemolysins. Med. Mycol. 51, 1–16.

Nguyen, T.H., Nagasaka, R., Ohshima, T., 2012. Effects of extraction solvents, cooking procedures and storage conditions on the contents of ergothioneine and phenolic compounds and antioxidative capacity of the cultivated mushroom *Flammulina velutipes*. Int. J. Food Sci. Technol. 47, 1193–1205.

Popović, V., Živković, J., Davidović, S., Stevanović, M., Stojković, D., 2013. Mycotherapy of cancer: an update on cytotoxic and antitumor activities of mushrooms, bioactive principles and molecular mechanisms of their action. Curr. Top. Med. Chem. 13, 2791–2806.

Ren, L., Perera, C., Hemar, Y., 2012. Antitumor activity of mushroom polysaccharides: a review. Food Funct. 3, 1118–1130.

Rob, O., Mlcek, J., Jurikova, T., 2009. Beta-glucans in higher fungi and their health effects. Nutr. Rev. 67, 624–631.

Roupas, P., Keogh, J., Noakes, M., Margetts, C., Taylor, P., 2012. The role of edible mushrooms in health: evaluation of the evidence. J. Funct. Foods 4, 687–709.

Ruthes, A.C., Smiderle, F.R., Iacomini, M., 2015. D-Glucans from edible mushrooms: a review on the extraction, purification and chemical characterization approaches. Carbohydr. Polym. 117, 753–761.

Sapozhnikova, Y., Byrdwell, W.C., Lobato, A., Romig, B., 2014. Effects of UV-B radiation levels on concentrations of phytosterols, ergothioneine, and polyphenolic compounds in mushroom powders used as dietary supplements. J. Agric. Food Chem. 62, 3034–3042.

Schwan, W.R., 2012. Mushrooms: an untapped reservoir for nutraceutical antibacterial applications and antibacterial compounds. Curr. Top. Nutraceutical Res. 10, 75–82.

Singh, R.S., Bhari, R., Kaur, H.P., 2010. Mushroom lectins: current status and future perspectives. Crit. Rev. Biotechnol. 30, 99–126.

Singh, S.S., Wang, H., Chan, Y.S., Pan, W., Dan, X., Yin, C.M., et al., 2015. Lectins from edible mushrooms. Molecules 20, 446–469.

Soares, A.A., de Souza, C.G.M., Daniel, F.M., Ferrari, G.P., da Costa, S.M.G., Peralta, R.M., 2009. Antioxidant activity and total phenolic content of *Agaricus brasiliensis* (*Agaricus blazei* Murill) in two stages of maturity. Food Chem. 112, 775–781.

Soler-Rivas, C., Ramírez-Anguiano, A.C., Reglero, G., Santoyo, S., 2009. Effect of cooking, in vitro digestion and Caco-2 cells absorption on the radical scavenging activities of edible mushrooms. Int. J. Food Sci. Technol. 44, 2189–2197.

Synytsya, A., Míčková, K., Synytsya, A., Jablonský, I., Spěváček, J., Erban, V., et al., 2009. Glucans from fruit bodies of cultivated mushrooms *Pleurotus ostreatus* and *Pleurotus eryngii*: structure and potential prebiotic activity. Carbohydr. Polym. 76, 548–556.

Tsai, S.Y., Tsai, H.L., Mau, J.L., 2008. Non-volatile taste components of *Agaricus blazei*, *Agrocybe cylindracea* and *Boletus edulis*. Food Chem. 107, 977–983.

Vamanu, E., Pelinescu, D., Avram, I., Nita, S., 2013. An in vitro evaluation of antioxidant and colonic microbial profile levels following mushroom consumption. BioMed Res. Int., 9 p. http://dx.doi.org/10.1155/2013/289821.

Varrot, A., Basheer, S.M., Imberty, A., 2013. Fungal lectins: structure, function and potential applications. Curr. Opin. Struct. Biol. 23, 678–685.

Wasser, S.P., 2002. Medicinal mushroom as a source of antitumor and immunomodulating polysaccharides. Appl. Microbiol. Biotechnol. 60, 258–274.

Weigand-Heller, A., Kris-Etherton, P.M., Beelman, R.B., 2012. The bioavailability of ergothioneine from mushrooms (*Agaricus bisporus*) and the acute effects on antioxidant capacity and biomarkers of inflammation. Prev. Med. 54, 575–578.

Woldegiorgis, A.Z., Abate, D., Haki, G.D., Ziegler, G.R., 2014. Antioxidant property of edible mushrooms collected from Ethiopia. Food Chem. 157, 30–36.

Xu, X., Yan, H., Chen, J., Zhang, X., 2011. Bioactive proteins from mushrooms. Biotechnol. Adv. 29, 667–674.

Yahaya, N.F.M., Rahman, M.A., Abdullah, N., 2014. Therapeutic potential of mushrooms in preventing and ameliorating hypertension. Trends Food Sci. Technol. 39, 104–115.

Yang, H., Hwang, I., Kim, S., Hong, E.J., Jeung, E.B., 2013. *Lentinus edodes* promotes fat removal in hypercholesterolemic mice. Exp. Ther. Med. 6, 1409–1413.

Zhang, Y., Li, S., Wang, X., Zhang, L., Cheung, P.C.K., 2011. Advances in lentinan: isolation, structure, chain conformation and bioactivities. Food Hydrocolloids 25, 196–206.

Zhang, N., Chen, H., Zhang, Y., Ma, L., Xu, X., 2013. Comparative studies on chemical parameters and antioxidant properties of stipes and caps of shiitake mushroom as affected by different drying methods. J. Sci. Food Agric. 93, 3107–3113.

Zied, D.C., Giménez, A.P., González, J.E.P., Dias, E.S., Carvalho, M.A., de Almeida Minhoni, M.T., 2014. Effect of cultivation practices on the β-glucan content of *Agaricus subrufescens* basidiocarps. J. Agric. Food Chem. 62, 41–49.

CHAPTER 5

Detrimental Compounds and Effects

Contents

As mentioned, this book does not deal with toxic or inedible mushroom species. Nevertheless, even some culinary species contain compounds that can be potentially injurious to human health.

5.1 POTENTIALLY PROCARCINOGENIC COMPOUNDS

In 1960 a natural compound, the chemical structure of which was suspected to be carcinogenic, was isolated from several

155

Agaricus species. This was subsequently proven. The compound was named agaritine. Isolation of several chemically related compounds followed. The second watched procarcinogen was isolated from *Gyromitra esculenta* and was designated gyromitrin. It also has several chemically related compounds.

The procarcinogenicity is caused by nitrogen–nitrogen bonds, and particularly by the derivatives of hydrazine NH_2–NH_2 or diazonium cation —$N^{(+)}$=N. An overview was published by Toth (1995). It is not yet clear why some mushroom species produce such compounds that are innocuous for them but potentially harmful for the consumers.

5.1.1 Agaritine

Agaritine [β-*N*-(γ-L-(+)-glutamyl)-4-(hydroxymethyl)-phenyl-hydrazine] is not a factual carcinogen, it is only a weakly mutagenic compound. Within organisms of mammals, however, it undergoes changes leading to the production of detrimental carcinogens. A simplified scheme of the changes is given in Fig. 5.1. Repeated feeding of experimental animals with either high doses of *Agaricus bisporus* or the isolated carcinogens caused tumor growth in several parts of the digestive tract and bladder.

Overall information has been published by Andersson and Gry (2004) and Roupas et al. (2010). Widely ranging agaritine contents in cultivated *A. bisporus* were reported, with usual levels being 200–500 mg kg^{-1} fresh matter (FM). The contents

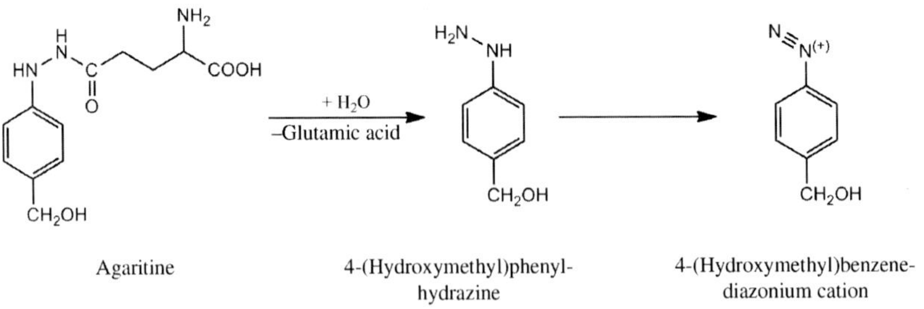

Figure 5.1 A simplified scheme of carcinogen formation from agaritine.

up to 6500 mg kg^{-1} were observed in dried *A. bisporus*. Agaritine is distributed unevenly within fruit bodies. The highest level was observed in the skin of caps and in the gills, with the lowest being in stipes (Schulzová et al., 2002).

In a survey of wild *Agaricus* species (Schulzová et al., 2009), 24 of 53 species contained more than 1000 mg kg^{-1} FM, and only 9 species contained less than 100 mg kg^{-1} FM. The highest content was up to 10,000 mg kg^{-1} FM and was determined in *Agaricus elvensis*. No correlation was observed between agaritine content and mushroom size, year, season, or site of collection. Moreover, agaritine was detected in some species of the genera *Leucoagaricus* and *Macrolepiota*.

A considerable decrease of agaritine content was observed during storage under refrigerator and freezer conditions, as well as during drying of white and brown *A. bisporus*, whereas freeze drying caused no loss. Boiling for 5 min degraded approximately one-fourth of agaritine, and approximately one-half has been extracted into the cooking broth. A further decrease was observed during prolonged boiling. Dry baking, similar to pizza production, reduced agaritine content by approximately one-quarter, whereas frying in oil or butter or deep frying resulted in a reduction of 35–70% of the initial level. Approximately two-thirds of agaritine were lost during microwave processing (Schulzová et al., 2002). In canned *A. bisporus*, usual agaritine levels were only approximately 15–20 mg kg^{-1} (ie, approximately 10-times lower than in fresh mushrooms) (Andersson et al., 1999).

Agaritine level in fresh *A. bisporus* was not affected by gamma irradiation up to 3 kGy, whereas at 5 kGy a significant reduction from 1540 to 1350 mg kg^{-1} dry matter (DM) occurred (Sommer et al., 2009).

It is not yet known for which products there are agaritine and other phenylhydrazine derivatives degraded during storage, preservation, and cooking. It is possible that such compounds can be biologically detrimental. However, Roupas et al. (2010) summed up that, based on the available evidence, agaritine consumption from cultivated *A. bisporus* poses no known toxicological risk to healthy humans.

Surprisingly, favorable health effects of agaritine were recently reported. It has been shown to be a potential anti–HIV agent and to exert antitumor activity against leukemia cells.

5.1.2 Gyromitrin

Species of genus *Gyromitra* are abundant and popular due to their delicious taste, particularly in the Nordic countries (Karlson-Stiber and Persson, 2003). Some of them are nontoxic, whereas others, including the widely distributed *G. esculenta* (false morel), contain toxic levels of gyromitrin (40–700 mg kg^{-1} FM) and several minor chemically related compounds.

Once ingested, gyromitrin (acetaldehyde-*N*-methyl-*N*-formyl-hydrazone or 2-ethylidene-1-methylhydrazide) is rapidly hydrolyzed under acidic conditions of the stomach, forming easily absorbed carcinogenic and toxic hydrazines (see Fig. 5.2). The individual response to gyromitrin is variable. Intoxications have occurred not only by eating *G. esculenta* but also by inhaling the vapors from cooking the mushroom. Typical toxic effects were observed in individuals who often handle gyromitrin-containing mushrooms.

After an acute, toxic exposure, vomiting and diarrhea occur after several hours of latency. The most typical features are disturbances of the central nervous system and blood glucose level. In severe intoxication, a second hepatorenal phase occurs after a period free of symptoms. These cases can be fatal.

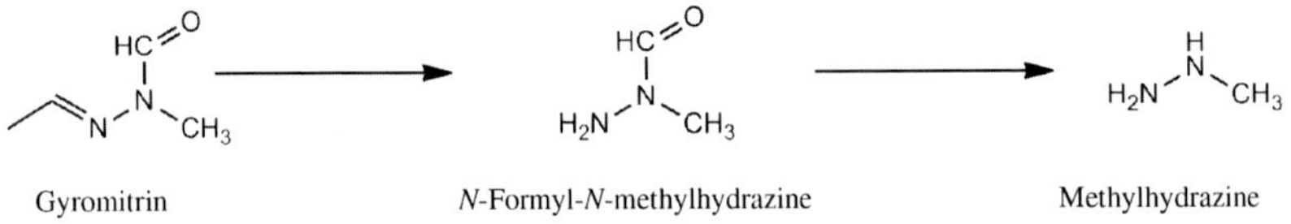

| Gyromitrin | *N*-Formyl-*N*-methylhydrazine | Methylhydrazine |

Figure 5.2 A scheme of toxic hydrazine formation from gyromitrin.

Gyromitrin content is reduced by prolonged air drying or by cooking of fresh as well as dried mushrooms. However, only free molecules of gyromitrin are removed during drying; for those that are organically bound, approximately half of the toxin content remains. Boiling must be repeated several times in large volumes of water in an open vessel to enable evaporation of volatile hydrazines. If cooked in a closed pot, the toxic substances would be retained.

5.2 FORMALDEHYDE

The initial information on the occurrence of formaldehyde in Chinese cultivated shiitake mushrooms (*Lentinula edodes*) in 2002 was proven in retail samples in the United Kingdom (Mason et al., 2004) and China (Liu et al., 2005). Mason et al. (2004) reported 11–24 mg, and Liu et al. (2005) reported 11.9–49.4 mg $100\,g^{-1}$ FM. The inceptive hypothesis supposed that the residues of formaldehyde were a result of disinfectant use during mushroom cultivation. However, samples of UK and Chinese origin, which were verified as being produced without any formaldehyde treatments, contained 10–32 mg $100\,g^{-1}$ FM (Mason et al., 2004).

Storage of shiitake mushrooms for 10 days had no effect on formaldehyde content, whereas frying for 6 min significantly reduced its level (Mason et al., 2004).

It seems plausible that formaldehyde in shiitake is of endogenous origin. It is probably released as one of numerous products of multistep hydrolysis of sulfur-containing lentinic acid, a precursor of compounds participating in characteristic aroma of shiitake (see Section 3.1.2).

The toxicological risk of formaldehyde intake from the consumed shiitake was assessed by Claeys et al. (2009). Formaldehyde is carcinogenic only via inhalation, not by ingestion. Moreover, the estimated daily exposure from shiitake consumption of $0.19\,\mu g\ kg^{-1}$ body weight in consumers appears to be very low as compared with total daily dietary intake of approximately

$100\,\mu g\ kg^{-1}$ body weight in the total population. Overall, formaldehyde in shiitake is not a cause for concern.

5.3 NICOTINE

Nicotine is the main alkaloid of tobacco leaves, with contents usually ranging between 0.3% and 3% of dry weight. Low levels also occur in other plants of the family Solanaceae, such as tomatoes, potatoes, peppers, and aubergines. Nicotine can cause dizziness, increased heart rate, and elevated blood pressure.

Surprisingly, an official German report on significant levels of nicotine in dried cepes ("true boletes"; *Boletus* spp.) from China in 2009 followed by reports on its occurrence in chanterelles (*Cantharellus cibarius*) raised questions regarding nicotine origin and possible human health concern. In 2009, the European Food Safety Authority proposed temporary nicotine residue levels of $0.04\,mg\ kg^{-1}$ for fresh cultivated and wild mushrooms and $1.2\,mg\ kg^{-1}$ for dry wild mushrooms, with the exception of dried cepes, which had a level of $2.3\,mg\ kg^{-1}$. These levels were prolonged (EU Commission Regulation No 897/2012).

Available literature deals primarily with the development of proper analytical methods for nicotine determination, whereas the data on nicotine content in mushrooms have been supplementary. Cavalieri et al. (2010) in Italy reported $0.01–0.04\,mg\ kg^{-1}$ in 10 fresh or frozen samples and $0.1–4.5\,mg\ kg^{-1}$ after drying. In Germany, approximately 30% of 79 analyzed samples of many wild species contained between 0.006 and $0.02\,mg\ kg^{-1}$ DM, and only approximately 7% contained more than $0.09\,mg\ kg^{-1}$ DM (Müller et al., 2011). However, nicotine contents in commercial dried cepes were in the range of $0.03–3.30\,mg\ kg^{-1}$ DM. Similarly in Spain (Lozano et al., 2012), all 20 analyzed commercial samples contained nicotine at low levels between 0.006 and $0.013\,mg\ kg^{-1}$ DM. Chinese data for several species vary within the range of $0.02–0.08\,mg\ kg^{-1}$ DM (Lin et al., 2013; Wang et al., 2011).

Until now, no clear reason has been established for this unexpected nicotine occurrence. Several factors have been considered, namely: the use of pesticides containing nicotine, which have been banned within the EU since 2010 but are still used in some countries; mushroom cultivation on contaminated substrate; cross-contamination under bad practices of processing and packaging (eg, drying of mushrooms in stoves used for tobacco drying); endogenous origin; and increased biosynthesis of nicotine via putrescine produced under the improper conditions during mushroom storage, handling, and drying (Schindler et al., 2015).

5.4 COPRINE

Fruit bodies of several species of the genus *Coprinus*, including plentiful *Coprinus comatus* and *Coprinus atramentarius*, contain nonprotein amino acid coprine, which is converted in the human body to toxic cyclopropanone hydrate (Fig. 5.3). Content of approximately $150\,mg$ coprine kg^{-1} FM is usual in European *C. atramentarius*. Low levels of coprine were also observed in *Clitocybe clavipes*, *Pholiota squarrosa*, and *Boletus erythropus*.

If alcohol is consumed up to approximately 72 h after the ingestion of species containing coprine, then alcohol intolerance occurs, similar to antabus effects, with an intense indisposition for several hours, including decreased blood pressure, elevated heart activity, flushing, nausea, vomiting, and headache. The cause of these symptoms is the blocking of alcohol detoxification (ethanol → acetaldehyde → acetic acid) by cyclopropanone hydrate in the second step and accumulation of detrimental acetaldehyde. This has been discussed in detail by Matthies and Laatsch (1992).

5.5 XENOBIOTICS

Prochloraz has been the most commonly used pesticide in mushroom production. Its residual levels in *A. bisporus* from seven Turkish facilities ranged widely from undetectable to $105.4\pm16.9\,\mu g\ kg^{-1}$ (Cengiz et al., 2014).

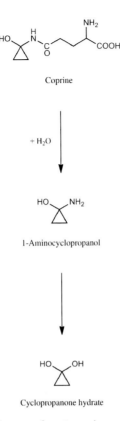

Figure 5.3 A simplified scheme of toxic cyclopropanone hydrate formation from coprine.

Residues of formerly used pesticides, such as chlorinated hydrocarbon DDT, its metabolites DDE and DDD, and γ-HCH (lindane, gammexane), were detected in all samples of *Boletus edulis* and *Xerocomus badius* from northeastern Poland. Mean sums of DDT were 3.78 and 1.71 µg kg^{-1} in fresh caps of *X. badius* and *B. edulis*, respectively, and respective contents of γ-HCH were 0.125 and 0.11 µg kg^{-1} (Gałgowska et al., 2012). Such levels do not exceed the acceptable levels and do not pose a health risk.

Thiabendazole counts among the most commonly used postharvest systemic fungicides for the control of fungal diseases during storage and distribution of fruits and vegetables, including mushrooms. Zhang et al. (2014) tested thiabendazole uptake and degradation by cultivated *Pleurotus eryngii*, *Pleurotus ostreatus*, and *Hypsizygus marmoreus*. The fungicide was either sprayed directly on substrates or premixed with the substrates. In the variants and in all three species, the highest residues were determined in the usually nonconsumed volva, which has been partially buried in substrate, and the lowest level was found in the stipe. The spraying resulted in higher thiabendazole residues than the premixing treatment. However, the residues were less than $5 \, mg \, kg^{-1}$, and the maximum residue limit in the EU has been $10 \, mg \, kg^{-1}$. The highest level was observed in *P. eryngii*. Half-lives of thiabendazole were 13.7, 13.6, and 10.0 days for *P. ostreatus*, *H. marmoreus*, and *P. eryngii*, respectively, in the variant with the premixed fungicide.

An insecticide fipronil is applied for the protection of sugarcane crops. Sugarcane bagasse is commonly used as a component of composts for the cultivation of *Agaricus subrufescens*. Bioaccumulation of fipronil from experimental composts (up to $32 \, mg \, kg^{-1}$) was not detected, whereas the fruit bodies accumulated the insecticide from the casing layer (Carvalho et al., 2014).

No detectable risky coplanar polychlorinated biphenyls (PCBs) were observed in cultivated *L. edodes*, *Flammulina velutipes*, and *Grifola frondosa* (Amakura et al., 2003). However, up to $13.1 \, \mu g \, kg^{-1}$ of lipids was reported by Kotlarska et al. (2010) in fresh wild mushrooms from northeastern Poland. Fruit bodies of cultivated mushrooms can be contaminated with PCBs from the growing substrate. However, only less than 0.1% of PCBs originally present in straw were transported into fruit bodies of *P. ostreatus* (Moeder et al., 2005). Similarly, no detectable contents of polychlorinated dibenzo-*p*-dioxins and negligible levels of polychlorinated dibenzofurans were reported by Amakura et al. (2003) in cultivated *L. edodes*, *F. velutipes*, and *G. frondosa*. Such levels were lower than in green leafy vegetables.

The trends in mycoremediation should be briefly mentioned here (for recent overviews see Kulshreshtha et al., 2014 and

Rhodes, 2014). Within "clean technologies," mycoremediation is based on the use of fungi and mushrooms for the removal of various pollutants from wastes, effluents, and soils. A variant combining remediation effects with the production of consumable fruit bodies is the most interesting from the point of view of this book.

Mycoremediation with mushrooms uses the processes of biodegradation, biosorption, and bioconversion. Biodegradation degrades and recycles complex molecules and some nonpolymeric recalcitrant pollutants to elementary mineral constituents such as water, carbon dioxide, nitrates, and others. Various mushroom species, including those widely cultivated (eg, *L. edodes*, *P. ostreatus*), produce numerous effective hydrolyzing and degrading enzymes for such processes. Biosorption is based on the sorption of metallic ions and organic pollutants from effluents by live or dried biosorbents such as mushroom mycelium and used mushroom compost. However, biosorption generates nonconsumable biomass.

On the contrary, bioconversion enables production of consumable fruit bodies of mushrooms cultivated on various agroindustrial and industrial wastes. Nevertheless, toxicity and genotoxicity of such fruit bodies have to be tested from case to case. Future research should be focused on the ability of a mushroom species to degrade pollutants in such a way that their disposal will not create another problem and fruit bodies can be consumed safely (Kulshreshtha et al., 2014).

5.6 NITRATES

The possible consequences of nitrates intake in humans, followed by their enzymatic reduction to nitrites, can be methemoglobinemia and the formation of carcinogenic nitrosamines. Plant foods and drinking water are the main sources of nitrates. Some vegetables such as spinach, lettuce, or radish can contain 2000–5000 mg kg^{-1} FM. The acceptable daily intake for nitrate anion (NO_3^-) was established as 3.65 mg kg^{-1} of body weight.

Only pilot information on nitrates in edible mushrooms has been available. Bóbics et al. (2015) analyzed 134 samples of 54 mushroom species (19, 13, 13, and 9 wild saprobic, ectomycorrhizal, wood-decaying, and cultivated, respectively). The highest mean nitrate contents were determined within saprobic species with a wide range of 151–12,720 mg kg^{-1} DM. The most accumulating species were *Clitocybe nebularis*, *Lepista nuda*, *Macrolepiota rhacodes*, and *Clitocybe odora*, with mean contents of 6980, 5840, 1880, and 1770 mg kg^{-1} DM, respectively. Considerably lower and relatively invariable mean contents of 217 and 229 mg kg^{-1} DM were observed in ectomycorrhizal and wood-decaying species, respectively. Within the cultivated species, *A. bisporus*, *P. ostreatus*, and *L. edodes* contained, on average, 563, 221, and 154 mg kg^{-1} DM, respectively.

Because the contents of nitrates expressed per FM are approximately 10-times lower than data per DM, nitrate contents in most of the analyzed mushroom species are comparable or lower than in nonaccumulating vegetables. Nevertheless, the body burden from a meal of 100 g of fresh *C. nebularis* or *L. nuda* would reach approximately one-quarter of the acceptable daily intake for a person with body weight of 70 kg.

5.7 RADIOACTIVITY

Both the economic and recreational collection of wild-growing mushrooms can be deprived or limited due to fears of mushroom contamination. Such a situation can arise both in sites heavily polluted with deleterious heavy metals (see Section 3.9) and in areas contaminated with radioactive fallout from a disaster at a nuclear facility.

Radionuclide contamination of the environment and food chains has originated from the fallout after nuclear weapons testing in the 1960s and the operation of nuclear energy–generating industries, including their accidents, and various applications of radioisotopes. Extreme situations occurred after the Chernobyl disaster in April 1986 (Ukraine) and the Fukushima disaster in March 2011 (Japan). Mushrooms are filamentous fungi and are very efficient at absorbing radionuclides from their substrate; they are an important component of long-term accumulation of radionuclides in upper horizons of forest soil due to the long-living and huge hyphal network and biomass.

Radioactive contamination of a great part of Europe after the Chernobyl accident triggered extensive research and monitoring of the environment, including mushrooms. Therefore, a lot of data on mushroom radioactivity have been available. The overall information is presented in several review articles focusing either on the environmental aspects and transfer, particularly of radiocesium to mushrooms (Duff and Ramsey, 2008; Gillett and Crout, 2000), or on the health aspects of radionuclide intake via the consumption of various mushroom species (Guillén and Baeza, 2014; Kalač, 2001, 2012). Only the weightiest original articles are cited, numerous references are cited in the mentioned reviews.

This section summarizes recent information on radioactivity of particular wild species from areas contaminated from the Chernobyl disaster. Data dealing with the repercussions of the Fukushima accident on mushroom contamination have been scarce. Cultivated mushrooms, usually growing indoors and on uncontaminated substrates, generally have considerably lower levels of anthropogenic radionuclides than wild species. A specific situation occurred recently in *L. edodes* grown outdoors on logs in open fields affected by the Fukushima fallout. Such fruit bodies were refrained from being sold.

5.7.1 Radioactivity Units and Legislation

One becquerel (Bq) is a unit for the activity of a radioactive source in which, on average, one atom decays per second. Activity concentration, for example, activity per weight unit (usually Bq kg^{-1}), has been commonly used. For the gravest radionuclides from the Chernobyl fallout (radiocesium $^{137}Cs + ^{134}Cs$), food statutory limits in the EU are 600 and 370 Bq kg^{-1} FM for adults and children, respectively. Thus, at usual mushroom DM of 10%, the limit for adults is equivalent to 6000 Bq kg^{-1} DM. The radioactivity of wild mushrooms in the period following the Chernobyl disaster incited legislative provisions and recommendations. The maximum permitted limit of the EU for ^{137}Cs has been 1250 Bq kg^{-1} FM (ie, 12,500 Bq kg^{-1} DM). Recently, as a consequence of the Fukushima disaster, a limit of 500 Bq kg^{-1} FM was established for vegetables, including mushrooms, in Japan.

5.7.2 Natural Radionuclides

As mentioned in Section 2.5, mushrooms contain considerably higher levels of potassium than do foods of plant origin. Potassium is distributed unevenly in fruit bodies. The content usually decreases in the following order: cap > stipe > gills or tubes > spores. The accumulation factor (ie, ratio of potassium content in fruit body DM and in DM of underlying substrate) is usually between 20 and 40. The radioisotope ^{40}K is present in the mixture of potassium isotopes at a constant level of 0.012%. Its half-life is extremely long.

The usual reported activity concentrations of ^{40}K in numerous wild-growing species have been in the range 800–1500 Bq kg^{-1} DM. It seems that the incorporation of stable potassium isotopes ^{39}K and ^{41}K, and hence also ^{40}K, from underlying soil to fruit bodies is self-regulated by the physiological requirements of a mushroom. Unlike ^{137}Cs from radioactive fallout, natural ^{40}K is distributed evenly in the vertical profile of forest soils. Cesium is chemically related to potassium; however, no significant correlation between ^{137}Cs and ^{40}K has been reported. Thus, different uptake mechanisms of these elements are suggested. The isotope ^{40}K is usually the main source of radioactivity in cultivated mushrooms.

Isotopes of lead ^{210}Pb (half-life 22.3 y) and polonium ^{210}Po (half-life 138 d) are decay products of naturally occurring parent uranium ^{238}U with intermediate isotopes radium ^{226}Ra and radon ^{222}Rn. These isotopes therefore occur at elevated levels in mushrooms in the vicinity of uranium mines. Ectomycorrhizal and saprobic species seem to have similar activity levels. The reported activity concentrations vary widely from <1 to hundreds of Bq kg^{-1} DM for each of the isotopes. Mushrooms take up ^{210}Pb directly from soils, and deposition from the atmosphere onto fruit bodies is limited. Dose coefficients (see Section 5.7.5) of ^{210}Po and ^{210}Pb are for approximately three or two orders of magnitude higher than that of ^{40}K, and the effective dose from their intake from mushrooms should not be overlooked. This particularly occurs in mushrooms from sites not contaminated with anthropogenic radionuclides.

Other natural radionuclides, uranium ^{238}U and thorium ^{232}Th, are taken up at lower levels than ^{40}K because they are not essential for mushroom development and have only a limited role in the effective dose of radioactivity. Similarly, low activity concentrations of radium ^{226}Ra, originating from decay of ^{238}U, were reported from France and in mushrooms from a Romanian uranium mining area. Beryllium isotope ^{7}Be (half-life 53.4 d) generated by the interaction of high-energy cosmic rays with the atmosphere can be deposited onto fruit bodies by precipitation. The health risk associated with this is very low.

5.7.3 Anthropogenic Radionuclides

The global environment was contaminated with several radionuclides from nuclear weapons testing until 1963. The total release of ^{137}Cs, the most important contaminant, was estimated as 9.6×10^{17} Bq. Limited data from up to 1985 have been available for mushroom radioactivity. In Central Europe, activity concentrations of ^{137}Cs were usually less than 1000 Bq kg^{-1} DM. X. badius and *Xerocomus chrysenteron* were, at that time, identified as accumulating species. Radionuclides of strontium ^{90}Sr and of plutonium were observed in wild mushrooms at toxicologically unimportant levels.

Mushroom contamination changed dramatically after the disaster at the Chernobyl nuclear power station on April 26, 1986. This disaster released into the environment approximately 3.8×10^{16} Bq from ^{137}Cs decay. Cesium ^{137}Cs and strontium ^{90}Sr were the main radionuclides produced during the explosive fission reaction. Both these radionuclides have long physical half-lives of 30.2 and 28.8 years, respectively. Radiocesium ^{134}Cs, with a half-life of 2.06 years, is produced in reactors during long-term fission. The ratio of ^{137}Cs to ^{134}Cs early after the disaster was approximately 2:1. Moreover, numerous further radionuclides of limited toxicological risk, namely ^{144}Ce, ^{131}I, ^{95}Nb, ^{239}Pu, ^{240}Pu, ^{103}Ru, ^{106}Ru, ^{230}Th, ^{232}Th, and ^{95}Zr, were detected in mushrooms soon after the accident.

Levels of local radioactivity contamination varied very widely within 10^1–10^5 Bq m^{-2}. The variability was affected by the direction and speed of radioactive clouds, distance from the Chernobyl power station, and particularly by dry or wet conditions in areas with rainfall. The levels of fallout were very different even in relatively adjacent sites. In the most heavily contaminated areas close to the power station, ^{137}Cs radioactivity may be a health concern for up to next three centuries. During the initial period after the disaster, both radiocesium isotopes participated in the radioactivity of the environment. Since approximately the mid-1990s, ^{137}Cs has remained the crucial radionuclide.

Wild edible mushrooms have been collected mostly from forests. Temperate forest soils are multilayer, comprising organic forest floor (debris), semi-organic, and mineral layers. The long-term retention of radionuclides in organic layers is considerably affected by fungal and microbial activities. Organic matter of forest soils has a lower affinity for cesium than mineral components. Thus, cesium is easily available for mushroom mycelium located in organic layers. Long-lasting availability of some radionuclides was therefore shown to be the source of their considerably higher transfer in forest ecosystems than in agricultural lands (Calmon et al., 2009).

5.7.4 Mushroom Radioactivity After the Chernobyl Disaster

The high variability of ^{137}Cs contamination, both spatial and temporal, was observed even within the same mushroom species in the years after the Chernobyl accident. Several factors have been implicated, such as mycelium habitat and depth, forest type, nutritional strategy of a species, soil clay content, soil moisture, and/or microclimate. In species with superficial mycelium, such as in genera *Clitocybe* and *Collybia*, the radioactivity of fruit bodies increased within a few months after the fallout deposition and gradually decreased. In deeper-penetrating species (eg, *B. edulis*), the contamination achieved peaks several years after the deposition (mostly in 1988, the second year after the Chernobyl disaster) and remained stable for a relatively long period. Thus, it has been difficult to estimate ecological half-lives. There exists a consensus that species of different nutritional strategy accumulate ^{137}Cs in the following order: ectomycorrhizal > saprobic > or ≈ parasitic (Gillett and Crout, 2000, and references therein). The highest accumulation in mycorrhizal species is attributed to the fact that the host plant can distinguish cesium from potassium, and thus the mushroom acts as a filter for the host plant. The transfer of ^{137}Cs to fruit bodies of mushrooms growing in coniferous forests usually exceeds the level in deciduous forests. The radioisotopes of cesium are distributed unevenly in the fruit bodies, with the following order: spore-bearing part (gills or tubes) > rest of the caps > stipes.

High accumulation of radiocesium in caps of *X. badius* and in related *B. erythropus* have been ascribed to natural polyphenolic pigments badion and norbadion A, causing a chocolate brown or golden yellow coloring. These pigments have numerous functional groups able to bind monovalent cations, including Cs^+.

Data on mushroom radioactivity until the late 1990s are reviewed by Gillet and Crout (2000), Kalač (2001), and Duff and Ramsey (2008). Numerous references are available therein. Generally, activity concentrations varied very widely, between a few hundred to more than $100,000\,Bq\ kg^{-1}$ DM in ectomycorrhizal species, and from a few hundred to a few thousand Bq kg^{-1} DM in saprobic and parasitic species. Thus, the EU statutory limit of $12,500\,Bq\ kg^{-1}$ DM was often extensively surpassed.

Table 5.1 Selected edible mushroom species with different rates of radiocesium ^{137}Cs accumulation

High	Medium
Cantharellus lutescens (M)	*Agaricus silvaticus* (S)
Cantharellus tubaeformis (M)	*Boletus edulis* (M)
Hydnum repandum (M)	*Cantharellus cibarius* (M)
Laccaria amethystea (M)	*Leccinum aurantiacum* (M)
Rozites caperata (M)	*Leccinum scabrum* (M)
Russula cyanoxantha (M)	*Russula xerampelina* (M)
Suillus luteus (M)	
Suillus variegatus (M)	
Xerocomus badius (M)	
Xerocomus chrysenteron (M)	

Nutritional strategy: M, ectomycorrhizal; S, saprobic.

Commonly consumed European species with high and medium levels of ^{137}Cs accumulation are provided in Table 5.1.

Very wide ranges exist even within a species. For instance, a unique study in Poland systematically covered the whole country in 1991. In total, 278 samples of highly accumulating and widely consumed *X. badius* were analyzed for ^{137}Cs, ^{134}Cs, and ^{40}K. The most frequent activity concentrations were 2000–10,000 and 200–600 Bq kg^{-1} DM for ^{137}Cs and ^{134}Cs, respectively. The highest levels, 157,000 and 16,300 Bq kg^{-1} DM for ^{137}Cs and ^{134}Cs, respectively, were observed in a "hot spot" site that was extremely contaminated by the Chernobyl fallout (Mietelski et al., 1994). Mushrooms from that site still showed high radioactivity in 2007.

Data on mushroom radioactivity after 2000 considerably decreased as compared to the previous period. Overall, mushroom contamination with ^{137}Cs remained relatively stable during 2001–2010. Considering the overall course of ^{137}Cs activity concentration in widely consumed wild mushrooms in the Czech Republic during the period 1987–2011, Škrkal et al. (2013) reported a limited decrease, but with great fluctuations. Two main factors, level of site contamination by the Chernobyl fallout and mushroom species, have to be taken into consideration (for the relevant literature see Kalač, 2012).

5.7.5 Radioactivity Burden from Mushroom Consumption

A possible risk of radioactivity to human health is expressed as the effective dose (E), given in millisieverts (mSv) per year. The acceptable yearly burden for an adult has been 1 mSv. Contribution to the yearly effective ingestion dose in an adult from mushroom consumption can be calculated as follows:

$$E = Y \times Z \times d_c,$$

where

Y = annual intake of mushrooms (kg DM per person)
Z = activity concentration (Bq kg^{-1} DM)
d_c = dose coefficient (conversion factor) defined as the dose received by an adult per unit intake of radioactivity.

The dose coefficients are 1.2×10^{-6}, 2.8×10^{-7}, 6.9×10^{-7}, 1.3×10^{-8}, 1.9×10^{-8}, 2.8×10^{-8}, and 6.2×10^{-9} Sv Bq^{-1} for ^{210}Po, ^{226}Ra, ^{210}Pb, ^{137}Cs, ^{134}Cs, ^{90}Sr, and ^{40}K, respectively. Polonium ^{210}Po is thus the most radiotoxic and potassium ^{40}K is the least radiotoxic within the seven main radionuclides. For instance, the effective dose of 1 mSv would be accomplished by the consumption of approximately 77,000 Bq of ^{137}Cs. Nevertheless, such an estimate of the effective dose represents the worst case scenario. It is calculated using raw mushrooms and complete availability of radionuclides by humans. Annual intake of mushrooms has been variable both between countries and individuals and ranges from tens of grams to more than 10 kg of fresh mushrooms. Very high consumption of wild-growing species has been traditional in Slavonic countries.

Thus, naturally occurring radionuclides ^{226}Ra, ^{210}Pb, ^{210}Po and anthropogenic radionuclides ^{134}Cs and ^{137}Cs need to be monitored to ensure the adequate protection of mushroom consumers (Guillén and Baeza, 2014).

The estimated internal dose due to wild mushroom consumption during the two decades following the Chernobyl disaster varied widely among Europe and within the individual countries. The highest values (in mSv per year) were reported from the zones heavily contaminated with the Chernobyl fallout, with up to 0.5 in some areas of Russia and Norway and between 0.01 and 0.1 in Ukraine, Poland, Czech Republic, and other Nordic countries. However, considerably lower levels were calculated in countries with low radioactive fallout and with low consumption of wild-growing mushrooms.

The activity concentrations of radioactive cesium can be effectively lowered during the culinary treatment of mushrooms. Decreases caused by washing can amount to 20%, and various cooking treatments can cause decreases of approximately 50–70% of the initial level. The effectiveness increases with treatments that break cell walls (eg, by freezing followed by soaking).

5.7.6 Radiocesium in Game-Feeding Mushrooms

Mushrooms have been a significant source of radionuclide contamination in animal tissue. In the years following the Chernobyl disaster, contamination was observed in wild animals, such as roe deer (*Capreolus capreolus*), red deer (*Cervus elaphus*), reindeer (*Rangifer tarandus*), and wild boar (*Sus scrofa*), and in grazing domestic goats and sheep. Generally, meat of red deer was less contaminated than that of roe deer or wild boar. Levels of radioactivity have often highly surpassed the EU limit of $600\,\mathrm{Bq\,kg}^{-1}$ FM for adults.

Subsequent trends of game meat radioactivity differed for ruminants and wild boar. In roe deer and red deer, ^{137}Cs activity concentrations are rather low and have steadily decreased with time since the Chernobyl fallout, with a temporary increase during mushroom season in autumn. On the contrary, meat of wild boar has retained high levels of radioactivity and has even increased during in some periods. This has been due to different habitat and dietary habits. Roe deer usually live in a relatively small territory, feeding mostly on less contaminated plants apart from during the mushroom season. Wild boar are omnivores that accept a variety of feed and travel long distances. The data from several Central European countries indicate an underground mushroom *Elaphomyces granulatus* (common name: deer truffles) as an important source of radioactivity in wild boar. The reported mean ^{137}Cs activity concentration of the mushroom from the period after 2000 was very high, approximately 6000 Bq kg^{-1} FM (ie, approximately 10-times more in DM). The level of ^{137}Cs in wild boar meat can be reduced considerably by repeated leaching with table salt solutions.

5.8 DETRIMENTAL EFFECTS OF *TRICHOLOMA EQUESTRE*

The yellow tricholoma (*Tricholoma equestre* or *Tricholoma flavovirens*), a wild species growing particularly in sandy pinewoods, was considered edible and tasty. However, several outbreaks of delayed severe rhabdomyolysis, which is fatal in some cases due to kidney failure, following the repeated consumption of the species occurred in France and Poland in approximately 2000. In Poland, the afflicted individuals consumed 100–400 g of the fresh mushroom during three to nine consecutive meals.

The results of experiments with laboratory mice proved myotoxic effects and also indicated cardiotoxic and hepatotoxic effects of *T. equestre* (Nieminen et al., 2008). The myotoxicity, indicated by elevated activity of plasma creatine kinase, was elicited in laboratory mice experimentally fed seven edible species, *Albatrellus ovinus*, *B. edulis*, *C. cibarius*, *Leccinum versipelle*, and three *Russula* species. Nevertheless, a high daily dose of 9 g of dried mushroom per kg body weight during 5-day exposure was tested. The results support the hypothesis that the reported toxic effects are not specific to *T. equestre*, but rather probably represent an unspecific response requiring individual sensitivity and a significant amount of ingested mushrooms to manifest itself (Nieminen et al., 2006).

Moreover, Yin et al. (2014) isolated 15 new triterpenoids from related edible *Tricholoma terreum*. Two saponaceolides of these, which are abundant in the species, displayed acute toxicity with medium lethal dose (LD50) values of 88.3 and 63.7 mg kg^{-1} body weight when administered orally to mice. Both compounds elevated activity of serum creatine kinase in mice, indicating myotoxicity.

5.9 ALLERGY AND ADVERSE DERMAL AND RESPIRATORY REACTIONS TO MUSHROOMS

Molds are well known as sources of allergens, and mushroom spores are abundant in many parts of the world. Seasonal peaks of the spores are observed in temperate zones during spring and autumn, which coincide with the major periods of fruit body formation.

An overview of various forms of mushroom allergies was published by Helbling et al. (2002). It is concluded that mushrooms are important sources of aeroallergens, and sensitization to various species is by far more frequent than previously expected. Basidiospores have been demonstrated to cause respiratory allergy and have also been suggested to trigger inflammatory skin eruptions in a subgroup of patients with atopic eczema. Many mushroom allergens are supposed to be cross-reactive, suggesting that mushrooms may harbor an unknown array of new allergens.

Flagellate erythema may arise in some individuals following raw or undercooked shiitake (*L. edodes*) consumption. The majority of shiitake dermatitis cases were described in Asian countries; however, there are recent reports from Europe and Brazil. The mechanism of dermatitis is thought to be toxic and due to lentinan, a thermolabile polysaccharide from the group of beta–glucans (see Section 4.2). Moreover, handling shiitake can cause occupational asthma in sensitive individuals (Pravettoni et al., 2014).

Case reports on respiratory allergic diseases in persons with extensive contact with mushrooms were published during the past several decades. A systematic 3-year study was performed by Tanaka et al. (2001) in a modern Japanese plant producing *H. marmoreus*. More than 90% of workers were sensitized to the spores, 40% quit because of the symptoms, and 5% developed hypersensitivity pneumonitis. Similar results were discovered in an Irish plant producing *A. bisporus*. In 67% of workers one or more respiratory symptoms occurred (Hayes and Rooney, 2014).

5.10 MICROBIAL LOAD AND SAFETY OF FRESH MUSHROOMS

Fresh mushrooms are an ideal medium for microbial growth due to high moisture content, high water activity, and neutral pH values. Together with their high initial microbial load, these characteristics are responsible for their very limited postharvest shelf life of only a few days. The genus *Pseudomonas* predominates and considerably participates in the deterioration. Moreover, fresh mushrooms rank among the potential carriers of pathogenic food-borne bacteria. The probability is relatively greater in wild mushrooms than in cultivated species. Under natural environment conditions, fruit bodies of the wild mushrooms are exposed to numerous animals and insects, some of which are fungivorous. For instance, the ability of *Listeria monocytogenes* and *Staphylococcus aureus* to survive on fresh mushrooms has been described.

A survey from Spain (Venturini et al., 2011) of a microbial load of seven bacterial groups, yeasts, and molds in 402 commercial samples of 8 cultivated and 14 wild mushroom species has provided interesting results. The total microbial counts ranged between 4.4 and 9.4 log cfu g^{-1} (the counts are expressed as decimal logarithm, eg, 4 means 10^4 organisms, 9 means 10^9, etc.; cfu = colony-forming units, ie, viable organisms). The genus *Pseudomonas* prevailed, with counts of 3.7–9.3 log cfu g^{-1}. No significant differences were detected between mean counts of wild and cultivated species in all the microbial groups studied. No pathogens were isolated from cultivated mushrooms.

However, 6.5% of wild mushrooms were positive for *L. monocytogenes*, with high occurrences in *Tuber indicum*, *Calocybe gambosa*, and *Hygrophorus limacinus*. *Yersinia enterocolitica* was detected in only four samples, and *Salmonella* spp., *Eschericia coli* O157:H7, and *S. aureus* were not detected at all.

Among 295 samples of *F. velutipes* from Chinese plants, the prevalence of *L. monocytogenes* was 18.6%. The contamination originated from the mycelium-scraping machinery in three of four surveyed plants (Chen et al., 2014).

From the decontamination and disinfestation points of view, electron beam irradiation is a prospective method that does not cause more pronounced effects on the nutritional profile of mushrooms. The assayed dose of 10 kGy is the highest that has been recommended (Fernandes et al., 2015).

REFERENCES

Amakura, Y., Tsutsumi, T., Sasaki, K., Maitani, T., 2003. Levels and congener distribution of PCDDs, PCDFs and co-PCBs in Japanese retail fresh and frozen vegetables. J. Food Hyg. Soc. Jpn. 44, 294–302.

Andersson, H.C., Gry, J. 2004. Phenylhydrazines in the Cultivated Mushroom (*Agaricus bisporus*). Occurrence, Biological Properties, Risk Assessment and Recommendations. Nordic Council of Ministers, Copenhagen, 123 pp.

Andersson, H.C., Hajšlová, J., Schulzová, V., Panovská, Z., Hájková, L., Gry, J., 1999. Agaritine content in processed foods containing the cultivated mushroom (*Agaricus bisporus*) on the Nordic and Czech market. Food Addit. Contam. 16, 439–446.

Bóbics, R., Krüzselyi, D., Vetter, J., 2015. Nitrate content in a collection of higher mushrooms. J. Sci. Food Agric. http://dx.doi.org/10.1002/jsfa.7108

Calmon, P., Thiry, Y., Zibold, G., Rantavaara, A., Fesenko, S., 2009. Transfer parameter values in temperate ecosystems: a review. J. Environ. Radioact. 100, 757–766.

Carvalho, M.A., Marques, S.C., Martos, E.T., Rigitano, R.L.O., Dias, E.S., 2014. Bioaccumulation of insecticide in *Agaricus subrufescens*. Hortic. Bras. 32, 159–162.

Cavalieri, C., Bolzoni, L., Bandini, M., 2010. Nicotine determination in mushrooms by LC-MS/MS with preliminary studies on the impact of drying on nicotine formation. Food Addit. Contam. 27, 473–477.

Cengiz, M.F., Catal, M., Erler, F., Bilgin, A.K., 2014. Rapid and sensitive determination of the prochloraz in the cultivated mushroom, *Agaricus bisporus* (Lange) Imbach. Anal. Methods 6, 1970–1976.

Chen, M.T., Wu, Q.P., Zhang, J.M., Guo, W.P., Wu, S., Yang, X.B., 2014. Prevalence and contamination patterns of *Listeria monocytogenes* in *Flammulina velutipes* plants. Foodborne Pathog. Dis. 11, 620–627.

Claeys, W., Vleminckx, C., Dubois, A., Huyghebaert, A., Hofte, M., Daenens, P., et al., 2009. Formaldehyde in cultivated mushrooms: a negligible risk for the consumer. Food Addit. Contam. A 26, 1265–1272.

Duff, M.C., Ramsey, M.L., 2008. Accumulation of radiocesium by mushrooms in the environment: a literature review. J. Environ. Radioact. 99, 912–932.

Fernandes, Â., Barreira, J.C.M., Antonio, A.L., Rafalski, A., Oliveira, M.B.P.P., Martins, A., et al., 2015. How does electron beam irradiation dose affect the chemical and antioxidant profiles of wild dried *Amanita* mushrooms? Food Chem. 182, 309–315.

Gałgowska, M., Pietrzak-Fiećko, R., Felkner-Poźniakowska, B., 2012. Assessment of the chlorinated hydrocarbons residues contamination in edible mushrooms from the North-Eastern part of Poland. Food Chem. Toxicol. 50, 4125–4129.

Gillett, A.G., Crout, N.M.J., 2000. A review of ^{137}Cs transfer to fungi and consequences for modelling environmental transfer. J. Environ. Radioact. 48, 95–121.

Guillén, J., Baeza, A., 2014. Radioactivity in mushrooms: a health hazard? Food Chem. 154, 14–25.

Hayes, J.P., Rooney, J., 2014. The prevalence of respiratory symptoms among mushroom workers in Ireland. Occup. Med. (London) 64, 533–538.

Helbling, A., Brander, K.A., Horner, W.E., Lehrer, S.B., 2002. Allergy to Basidiomycetes In: Breitenbach, M. Crameri, R. Lehrer, S.B. (Eds.), Fungal Allergy and Pathogenicity. Chemical Immunology, vol. 81 Karger, Basel, pp. 28–47.

Kalač, P., 2001. A review of edible mushroom radioactivity. Food Chem. 75, 29–35.

Kalač, P., 2012. Radioactivity of European wild growing edible mushrooms. In: Andres, S., Baumann, N. (Eds.), Mushrooms: Types, Properties and Nutrition Nova Sci. Publ., New York, pp. 215–230.

Karlson-Stiber, C., Persson, H., 2003. Cytotoxic fungi—an overview. Toxicon 42, 339–349.

Kotlarska, M.M., Pietrzak-Fiećko, R., Smoczyński, S.S., Borejszo, Z., 2010. [The level of polychlorinated biphenyls in mushrooms available at the market in the region of Warmia and Masuria.] Żywność—Nauka, Technologia, Jakość 17, 49–57. (in Polish).

Kulshreshtha, S., Mathur, N., Bhatnagar, P., 2014. Mushroom as a product and their role in mycoremediation. AMB Express 4, 29. http://www.amb-express.com/content/4/1/29.

Lin, X., Sun, Y., Xu, D., Li, Y., Liu, S., Xie, Z., 2013. Sensitive capillary electrophoretic profiling of nicotine and nornicotine in mushrooms with amperometric detection. Electrophoresis 34, 2033–2040.

Liu, J.F., Peng, J.F., Chi, Y.G., Jiang, G.B., 2005. Determination of formaldehyde in shiitake mushroom by ionic liquid-base liquid-phase microextraction coupled with liquid chromatography. Talanta 65, 705–709.

Lozano, A., Martínez-Uroz, M.A., Gómez-Ramos, M.J., Gómez-Ramos, M.M., Mezcua, M., Fernández-Alba, A.R., 2012. Determination of nicotine in mushrooms by various GC/MS- and LC/MS-based methods. Anal. Bioanal. Chem. 402, 935–943.

Mason, D.J., Sykes, M.D., Panton, S.W., Rippon, E.H., 2004. Determination of naturally-occurring formaldehyde in raw and cooked shiitake mushrooms by spectrophotometry and liquid chromatography-mass spectrometry. Food Addit. Contam. 21, 1071–1082.

Matthies, L., Laatsch, H., 1992. [Unusual mushroom intoxication: coprine, emetic compound of alcohol intolerance.] Pharm. Unserer Zeit 21, 14–20. (in German).

Mietelski, J.W., Jasińska, M., Kubica, B., Kozak, K., Macharski, P., 1994. Radioactive contamination of Polish mushrooms. Sci. Total Environ. 157, 217–226.

Moeder, M., Cajthaml, T., Koeller, G., Erbanová, P., Šašek, V., 2005. Structure selectivity in degradation and translocation of polychlorinated biphenyls (Delor 103) with a *Pleurotus ostreatus* (oyster mushroom) culture. Chemosphere 61, 1370–1378.

Müller, C., Bracher, F., Plössl, F., 2011. Determination of nicotine in dried mushrooms by using a modified QuEChERS approach and GC-MS-MS. Chromatographia 73, 807–811.

Nieminen, P., Kirsi, M., Mustonen, A.-M., 2006. Suspected myotoxicity of edible wild mushrooms. Exp. Biol. Med. 231, 221–228.

Nieminen, P., Kärjä, V., Mustonen, A.-M., 2008. Indications of hepatic and cardiac toxicity caused by subchronic *Tricholoma flavovirens* consumption. Food Chem. Toxicol. 46, 781–786.

Pravettoni, V., Primavesi, L., Piantanida, M., 2014. Shiitake mushroom (*Lentinus edodes*): a poorly known allergen in Western countries responsible for severe work-related asthma. Int. J. Occup. Med. Environ. Health 27, 871–874.

Rhodes, C.J., 2014. Mycoremediation (bioremediation with fungi)—growing mushrooms to clean the earth. Chem. Speciation Bioavailability 26, 196–198.

Roupas, P., Keogh, J., Noakes, M., Margetts, C., Taylor, P., 2010. Mushrooms and agaritine: a mini-review. J. Funct. Foods 2, 91–98.

Schindler, B.K., Bruns, S., Lach, G., 2015. Biogenic amines—a possible source for nicotine in mushrooms? A discussion of published literature data. Food Chem. 171, 379–381.

Schulzová, V., Hajšlová, J., Peroutka, R., Gry, J., Andersson, H.C., 2002. Influence of storage and household processing on the agaritine content of the cultivated *Agaricus* mushroom. Food Addit. Contam. 19, 853–862.

Schulzová, V., Hajšlová, J., Peroutka, R., Hlaváček, J., Gry, J., Andersson, H.C., 2009. Agaritine content of 53 *Agaricus* species collected from nature. Food Addit. Contam. 26, 82–93.

Škrkal, J., Rulík, P., Fantínová, K., Burianová, J., Helebrant, K., 2013. Long-term ^{137}Cs activity monitoring of mushrooms in forest ecosystems of the Czech Republic. Radiat. Prot. Dosimetry 157, 579–584.

Sommer, I., Schwartz, H., Solar, S., Sontag, G., 2009. Effect of γ-irradiation on agaritine, γ-glutamyl-4-hydroxybenzene (GHB), antioxidant capacity, and total phenolic content of mushrooms (*Agaricus bisporus*). J. Agric. Food Chem. 57, 5790–5794.

Tanaka, H., Saikai, T., Sugawara, H., Tsunematsu, K., Takeya, I., Koba, H., et al., 2001. Three-year follow-up study of allergy in workers in a mushroom factory. Respir. Med. 95, 943–948.

Toth, B., 1995. Mushroom toxins and cancer (Review). Int. J. Oncol. 6, 137–145.

Venturini, M.E., Reyes, J.E., Rivera, C.S., Oria, R., Blanco, D., 2011. Microbiological quality and safety of fresh cultivated and wild mushrooms commercialized in Spain. Food Microbiol. 28, 1492–1498.

Wang, H., Zhao, Q., Song, W., Xu, Y., Zhang, X., Zeng, Q., et al., 2011. High-throughput dynamic microwave-assisted extraction on-line coupled with solid-phase extraction for analysis of nicotine in mushrooms. Talanta 85, 743–748.

Yin, X., Feng, T., Shang, J.H., Zhao, Y.L., Wang, F., Li, Z.H., et al., 2014. Chemical and toxicological investigation of a previously unknown poisonous European mushroom *Tricholoma terreum*. Chem.—Eur. J. 20, 7001–7009.

Zhang, Z., Jiang, W., Jian, Q., Song, W., Zheng, Z., Ke, C., et al., 2014. Thiabendazole uptake in shimeji, king oyster, and oyster mushrooms and its persistence in sterile and nonsterile substrates. J. Agric. Food Chem. 62, 1221–1226.

CHAPTER 6

Conclusions

Contents

Both cultivated and wild-growing edible mushrooms are consumed as a delicacy, because of their characteristic smell, taste, and texture, or as a healthy food, because of their low calories and high fiber content. Recent global production of approximately 20 cultivated species has exceeded 10 million tons and is rapidly increasing, with China being the leading producer. More than 2000 wild species are safe for consumption. The number of wild species collected and consumed in various regions of the world is unknown; however, it is probably hundreds of species.

Recent information on composition, proteins, carbohydrates, lipids, vitamins, minerals, numerous minor components, and health-stimulating and adverse compounds of culinary mushrooms is presented in the previous chapters. However, the composition of toxic, inedible, and medicinal species is not included.

Available nutritional data deal mostly with raw mushrooms, whereas information on the changes in nutrients during various preservation, culinary, and processing treatments and storage has been limited. Even more scarce are data on bioavailability. High levels of indigestible polysaccharides, valued as dietary fiber, most probably decrease the digestibility of the nutrients.

6.1 PROXIMAL COMPOSITION AND NUTRIENTS

Dry matter (DM) of mushrooms is low, usually $8–14\,\mathrm{g}$ $100\,\mathrm{g}^{-1}$ fresh matter (FM). DM $10\,\mathrm{g}$ per $100\,\mathrm{g}$ FM (10%) has been commonly used for the conversion between DM and FM if the actual DM is unknown. Usual proximal compositions are 20–25, 2–3, and $5–12\,\mathrm{g}$ $100\,\mathrm{g}^{-1}$ DM for crude protein, crude fat, and ash (minerals), respectively. Various carbohydrates form the rest. Due to very low DM and fat content, mushrooms are a low-energy food item. The calculated energy value mostly ranges between 300 and $400\,\mathrm{kcal}$ $(1250–1670\,\mathrm{kJ})$ kg^{-1} FM. Nevertheless, such data are overestimated because a considerable part of polysaccharides is indigestible.

Previous data on protein content were overestimated by nearly one-third. Nevertheless, the nutritional value of mushroom protein seems to be higher as compared with most plant proteins. Amino acid composition varies among mushroom species. The proportion of essential (indispensable) amino acids varies, with approximately 40% of total amino acid content in wild species and between 30% and 35% in cultivated mushrooms. Methionine appears to be the very limiting essential amino acid.

The fat (lipids) content is very low. Great differences occur among species. Generally, unsaturated fatty acids, particularly linoleic acid (ω-6) and oleic acid, are prevalent. Palmitic acid is the most present saturated fatty acid. The proportion of valued ω-3 polyunsaturated fatty acids is very low. Therefore, mushrooms rank among food items with marginal nutritional lipid roles.

Mushrooms contain various carbohydrates as the main component of DM. Disaccharide α,α-trehalose and alcoholic sugar mannitol prevail among soluble sugars, usually at level of several grams per $100\,\mathrm{g}$ DM. Within polysaccharides, glycogen forms energy reserves, and nitrogen-containing chitin is the predominant component of cell walls and the main part of dietary fiber. Limited literature data on dietary fiber content vary. The level of approximately $25–30\,\mathrm{g}$ $100\,\mathrm{g}^{-1}$ DM with approximately half being insoluble fiber seems to be probable.

Ash content is formed by potassium and phosphorus, with usual levels of 2–4 and 0.5–1 g $100 g^{-1}$ DM, respectively. However, the sodium and calcium contents are very low. Generally, ash content is higher than or comparable to that of most vegetables.

Widely disseminated information about high vitamin levels of mushrooms, particularly B vitamins, has to be corrected. Mushrooms appear to be a good source of ergosterol, the precursor of vitamin D_2 (ergocalciferol), and vitamin B_{12}. However, the contents of other vitamins are comparable or lower than that of many vegetables.

6.2 MINOR CONSTITUENTS

The taste of mushrooms results from a combination of numerous nonvolatile constituents, particularly soluble sugars and polyols, free amino acids, 5′-nucleotides, and carboxylic acids. Great interest is focused on the constituents contributing to umami taste. The valued flavor, which is strengthened in dried mushrooms, is affected by many volatile compounds of various chemical natures (alcohols, aldehydes, ketones, terpenes, acids and their esters, aromatic, heterocyclic, and sulfur compounds). Among them, eight-carbon aliphatic constituents formed by a specific oxidation of linoleic acid are of extraordinary interest.

Mushroom pigment composition differs from that of plants. Chlorophylls and anthocyanins are lacking. Betalains, carotenoids, and other terpenoids occur only in some mushroom species. Quinones or similar conjugated structures form the great proportion of mushroom pigments. They are produced by enzymatic oxidation of polyphenols following mechanical damage of mushroom tissues. Such changes cause great economic losses, particularly due to browning of the white button mushroom, *Agaricus bisporus*.

The total content of aliphatic acids is up to several grams per 100 g DM. Malic, citric, and oxalic acids are prevalent. Some recent data reported a high content of oxalic acid in some species.

Individuals with kidney disorders, gout, osteoporosis, or rheuma-toid arthritis should limit the intake of such species.

Numerous phenolics of various chemical structures are the main sources of antioxidant activity. The usual total phenolic con-tents range between 0.1 and 0.6 g of gallic acid equivalents per 100 g DM. Phenolics are prone to oxidation. Among mushroom phenolics, the main interest has been on phenolic acids.

Several tens of sterols besides ergosterol, a provitamin form of D_2, were observed in some mushroom species. Existing research has focused primarily on medicinal species. The contents are lower by one to three orders of magnitude than those of ergos-terol. Until now, biological effects of the sterols have not been elucidated.

Overall, mushrooms seem to be rich in indole compounds, particularly in serotonin. Due to their observed instability dur-ing boiling, great changes in their profile can be expected during other thermal treatments. Mushrooms thus seem to be a promis-ing source of precursors of indole derivatives possessing various physiological activities in humans.

According to limited data, total purine contents in cultivated mushrooms seem to be acceptable for patients with gout or hyperuremia.

Among biogenic amines, which may indicate protein decom-position and matrix deterioration, putrescine is prevalent in wild species after harvest. Nevertheless, histamine and tyramine, the amines with the most undesirable health effects, occurred at minor levels. However, during mushroom storage, even under refrigeration, considerable formation of putrescine and cadaverine was observed. Raw mushrooms belong among food items with a high level of polyamine spermidine.

Data on trace element contents have been the most reported information regarding mushroom composition since the 1970s. Usual contents for most wild species grown in unpolluted sites are 50–200 (Al), <1 (As), 0.5–5 (Cd), 0.5–5 (Cr), 20–70 (Cu), 30–150

(Fe), <0.5–5 (Hg), 0.5–5 (Ni), 1–5 (Pb), 1–5 (Se), and 30–150 (Zn) mg kg^{-1} DM. Less numerous data are available for many other trace elements. These values can be increased, even by an order of magnitude, in mushrooms picked in polluted areas. Moreover, some species have accumulating and even hyper-accumulating abilities with various elements. Under such circumstances, wild mushrooms can be a health risk. The contents of trace elements in cultivated mushrooms are generally considerably lower than in wild-growing species.

Regarding other nutrients, most available data deal with fresh mushrooms, whereas information on the individual minor constituents during storage, processing, or cooking conditions has been limited, as have the data regarding bioavailability.

6.3 HEALTH-STIMULATING COMPOUNDS AND EFFECTS

Culinary mushrooms, in addition to medicinal species, have become a field of interest when searching for health-stimulating constituents. A group of so-called culinary–medicinal mushrooms has originated.

The antioxidant activity of many mushroom species is comparable with various fruits and vegetables. Phenolics are the most effective constituents. Drying decreases the antioxidant activity. Some mushroom polysaccharides, particularly β-glucans, exhibit antitumor, immunostimulating, and prebiotic properties. High-molecular-weight β-glucans are more effective than low-molecular-weight ones. Beta-glucans occur in fruit bodies at a level of hundreds to thousands of mg per 100 g DM. Researchers concentrate particularly on medicinal species. Various curative powers have been demonstrated for mushroom lectins, hemolysins, lovastatin, eritadenine, gamma-aminobutyric acid, and ergothioneine.

Overall, culinary mushrooms are regarded as the so far undiscovered pool of various compounds with beneficial health effects.

6.4 DETRIMENTAL COMPOUNDS AND EFFECTS

Even some edible mushroom species contain injurious compounds.

Potentially procarcinogenic agaritine occurs in the genus *Agaricus* and gyromitrin occurs in the genus *Gyromitra*, particularly in delicious *Gyromitra esculenta*. Both compounds chemically belong to hydrazine derivatives. Based on the available evidence, agaritine consumption from cultivated *A. bisporus* poses no toxicological risk to healthy humans. Drying or extensive cooking decreases the risk of gyromitrin toxicity.

Formaldehyde detected in *Lentinula edodes* (shiitake) most probably originates from lentinic acid, a dipeptide, and is not a cause for concern. No clear reason has been established for the surprising occurrence of nicotine at levels up to several mg per kg DM in some commercial mushrooms. Coprine is a nonprotein amino acid occurring in the genus *Comatus*. A decomposition product of coprine induces alcohol intolerance similar to antabus. Some saprobic species can accumulate considerable levels of nitrates. Mushrooms can contain residues of various xenobiotics, particularly pesticides. However, the data have been limited and dispersed.

Mushroom radioactivity has been extensively investigated, particularly during the decade following the Chernobyl nuclear power station disaster in 1986. Some ectomycorrhizal species, including those highly collected and consumed, have a high ability to accumulate radiocesium ^{134}Cs and ^{137}Cs. Such mushrooms are food items with the maximum observed radioactivity. The natural radioactivity of mushrooms is higher than that of other vegetables due to high potassium content, including radioactive isotope ^{40}K.

Tasty *Tricholoma equestre* and probably some other species can sporadically cause myotoxic, cardiotoxic, and hepatotoxic effects in sensitive consumers. Allergies and adverse dermal and respiratory reactions are reported in sensitive individuals after prolonged handling of several cultivated species. Wild mushrooms can be contaminated with pathogenic bacteria.

APPENDIX I

List of Abbreviations

Abbreviation	Expansion	See page
AAS	Atomic absorption spectroscopy	111
BCF	Bioconcentration factor	113
cfu	Colony forming unit	176
CLA	Conjugated linoleic acid	32
DDD	Dichlorodiphenyldichloroethane	162
DDE	Dichlorodiphenyltrichloroethene	162
DDT	Dichlorodiphenyltrichloroethane	162
DM	Dry matter (or dry weight)	17
DNA	Deoxyribonucleic acid	138
DOPA	3,4-Dihydroxyphenylalanine	88
EU	European Union	110
EUC	Equivalent umami concentration	80
FA	Fatty acid	22
FAO	Food and Agriculture Organization	21
FM	Fresh matter (or fresh weight)	17
GABA	Gamma-aminobutyric acid	147
GAE	Gallic acid equivalent	90
HCH	Hexachlorocyclohexane	162
ICP-AES	Inductively coupled plasma–atomic emission spectroscopy	122
IU	International unit (in vitamins)	56
MSG	Monosodium glutamate	76
MUFA	Monounsaturated fatty acid	24, 31, 191
PCB	Polychlorinated biphenyl	163
PCDD	Polychlorinated dibenzo-p-dioxin	177
PCDF	Polychlorinated dibenzofuran	177
PTMI	Provisional tolerable monthly intake	112
PTWI	Provisional tolerable weekly intake	112
PUFA	Polyunsaturated fatty acid	24, 31, 191
RNS	Reactive nitrogen substances	138
ROS	Reactive oxygen substances	138
SFA	Saturated fatty acid	24, 31, 191
TDF	Total dietary fiber	42
TPC	Total phenolic content	90
UK	United Kingdom	33
WHO	World Health Organization	112
UV	Ultraviolet light	55

APPENDIX II

Commonly Used Japanese Names of Mushrooms

Japanese common name (in English)	Scientific name
Agitake	*Pleurotus eryngii* var. *ferulae*
Aragekikurage	*Auricularia polytricha*
Bunashimeji	*Hypsizygus marmoreus*
Enokitake	*Flammulina velutipes*
Eringi	*Pleurotus eryngii*
Hanabiratake	*Sparassis crispa*
Hatakeshimeji	*Lyophyllum decastes*
Himarayahiratake	*Pleurotus sajor-caju*
Himematsutake	*Agaricus subrufescens* (syn. *Agaricus brasiliensis* or *Agaricus blazei*)
Hiratake	*Pleurotus ostreatus*
Houbitake	*Pleurotus sajor-caju*
Honshimeji	*Lyophyllum shimeji*
Kawaratake	*Trametes versicolor*
Kikurage	*Auricularia auricula-judae*
Maitake	*Grifola frondosa*
Mannentake	*Ganoderma lucidum*
Nameko	*Pholiota nameko*
Reishi	*Ganoderma lucidum*
Shiitake	*Lentinula edodes*
Shimeji	*Hypsizygus tessulatus*
Shirokikurage	*Tremella fuciformis*
Tokiirohiratake	*Pleurotus salmoneostramineus*
Tsukuritake	*Agaricus bisporus*
Yamabushitake	*Hericium erinaceus*

APPENDIX III

Characteristics of the Main Fatty Acids Occurring in Mushroom Lipids

Acid name	Symbol[a]	Position of double bond(s)[b]
Saturated fatty acids (SFA)		
Lauric	$C_{12:0}$	—
Myristic	$C_{14:0}$	—
Palmitic	$C_{16:0}$	—
Stearic	$C_{18:0}$	
Monounsaturated cis-fatty acids (MUFA)		
Palmitoleic	$C_{16:1n\text{-}7}$	9
Oleic	$C_{18:1n\text{-}9}$	9
Polyunsaturated cis-fatty acids (PUFA)		
Linoleic	$C_{18:2n\text{-}6}$	9,12
α-Linolenic	$C_{18:3n\text{-}3}$	9,12,15
Conjugated linoleic acid (CLA)		
Rumenic acid		9-*cis*, 11-*trans*

[a]X:Yn–Z X, number of carbon atoms; Y, number of double bonds; n–Z, position of double bond close to methyl carbon of an acid (also expressed as ω–Z).
[b]Positions of double bonds numbered from carboxylic carbon.

INDEX OF MUSHROOM SPECIES

Note: Page numbers followed by "*f*" and "*t*" refer to figures and tables, respectively.

SUBJECT INDEX

Note: Page numbers followed by "*f*" and "*t*" refer to figures and tables, respectively.

Photo 1 *Agaricus bisporus* white *Source: https://commons.wikimedia.org/ wiki/Category:Edible_mushrooms*

Photo 2 *Agaricus bisporus* brown *Source: https://commons.wikimedia.org/ wiki/Category:Edible_mushrooms*

Photo 3 *Agaricus subrufescens Source: Dr. Ivan Jablonský*

Photo 4 *Boletus edulis Source: https://commons.wikimedia.org/wiki/Category: Edible_mushrooms*

Photo 5 *Cantharellus cibarius Source: https://commons.wikimedia.org/wiki/ Category:Edible_mushrooms*

Photo 6 *Craterellus cornucupioides Source: Dr. Jan Borovička*

Photo 7 *Flammulina velutipes Source: Dr. Ivan Jablonský*

Photo 8 *Gyromitra esculenta Source: Dr. Jan Borovička*

Photo 9 *Lentinula edodes Source: Dr. Ivan Jablonský*

Photo 10 *Macrolepiota procera Source: Dr. Eva Dadáková*

Photo 11 *Pholiota nameko Source: Dr. Ivan Jablonský*

Photo 12 *Pleurotus eryngii Source: Dr. Ivan Jablonský*

Photo 13 *Pleurotus ostreatus Source: Dr. Jan Borovička*

Photo 14 *Sparassis crispa Source: Eva Dadáková*

Photo 15 *Tricholoma flavovirens Source: Dr. Jan Borovička*

Photo 16 *Xerocomus badius Source: Dr. Jan Borovička*